CHINESE SCIENCE AND TECHNOLOGY INDUSTRIAL PARKS

Chinese Science and Technology Industrial Parks

SUSAN M. WALCOTT
Georgia State University

ASHGATE

Published by
Ashgate Publishing Limited
Gower House
Croft Road
Aldershot
Hampshire GU11 3HR
England

Ashgate Publishing Company
Suite 420
101 Cherry Street
Burlington, VT 05401-4405
USA

Ashgate website: http://www.ashgate.com

British Library Cataloguing in Publication Data
Walcott, Susan M.
Chinese science and technology industrial parks
1. Industrial districts - China 2. High technology industries - China 3. Technology transfer - China 4. Investments, Foreign - China 5. China - Economic conditions - 1976-2000
6. China - Economic conditions - 2000-
I. Title
338.6'042'0951

Library of Congress Cataloging-in-Publication Data
Walcott, Susan, M. 1949-
Chinese science and technology industrial parks / Susan M. Walcott.
p. cm.
Includes bibliographical references and index.
ISBN 0-7546-0952-9 (alk.paper)
1. Research parks. 2. Research parks--China. I. Title.

T175.7.W35 2003
607'.251--dc21

2003041918

ISBN 0 7546 0952 9

Printed in Great Britain by Antony Rowe Ltd, Chippenham, Wiltshire

Contents

List of Figures

List of Tables

Acknowledgments

The overseas research underlying this book was financed through two grants: an exploratory Research Initiation Grant from Georgia State University, and a Board of Regents Global Partnerships Initiative fund. Their support extended through four summers of visits to and interviews with individuals in Beijing, Dongguan (during the International Geographical Union's 'Industrial Geography' conference in 2000), Shenzhen, Suzhou, Xi'an, and especially Shanghai.

Collaboration with Chinese colleagues opened doors in ways that are vital to this type of research. Professor Ning Yuemin of East China Normal University in Shanghai and Professor Wang Jici of Peking University in Beijing deserve particular thanks, as do several of their diligent graduate students who also assisted. Thanks also go to the numerous Chinese and Western business people and development park managers who gave their time for interviews and tours. Cartographic work by Jeffrey McMichael at Georgia State University was excellent and exacting, as always.

Susan M. Walcott, Department of Geography, Georgia State University

List of Abbreviations

ATP	Advanced Technology Products
BDA	Beijing Development Area
BEZ	Beijing Experimental Zone for the Development of New Technology Industries
CAS	Chinese Academy of Sciences
CSSIP	China-Singapore Suzhou Industrial Park
ETDZ	Economic Technology Development Zone
FDI	Foreign Direct Investment
HTP	High Tech Park (*gau keji yuan*)
IT	Information Technology
LDC	Less Developed Country
MDC	Most Developed Country
MNC	Multinational Corporation
MNE	Multinational Enterprise
MOFTEC	Ministry of Finance, Technology, Economic Cooperation
NID	New Industrial District
OECD	Organization of European Cooperative Development
PKU	Peking (Beijing) University (retaining the old name in the Wade-Giles romanization prevailing in the pre-1949 days of its founding)
PLA	People's Liberation Army
PRC	People's Republic of China
QU	Qinghua University, also known as Tsinghua University (retaining the old name in the Wade-Giles romanization prevailing in the pre-1949 days of its founding)
R&D	Research and Development
RMB	Renminbi, Chinese currency
SHIP	Shenzhen Hi Tech Industrial Park
SMETDZ	Shanghai Minhang Economic and Technological Development Zone
SOE	State Owned Enterprise
STC	Science and Technology Commission
STIP	Science and Technology Industrial Park
TNC	Transnational Corporation

TVE	Township-Village Enterprise
UAE	University-Affiliated Entity
WFOE	Wholly Foreign Owned Enterprise

Chapter 1

Chinese Economic Development Zones: An Overview

Introduction

China functions as the linchpin for Asia's future development. Issues of geography provide a crucial lens for examining economic prospects in specific places within this varied country roughly the size of the contiguous United States. The central concern of this book lies with the role of proximity among companies for promoting learning, enabling adaptation to new conditions. This includes challenges faced by both Chinese firms seeking to upgrade product and production techniques, and by foreign firms seeking a foothold in China's domestic market as well as competitive advantage for export production. A major finding of many place-based studies of successful businesses is the key importance of frequent interactions, creating 'learning districts' based on information exchange, which in turn enable companies to be more flexible and innovative. Chinese business parks featuring companies classified as possessing relatively advanced technology serve as spatially distinct areas set aside within large urban areas for the promotion of technology development through transfers from nearby foreign companies and domestic research centers. Comparing how this strategy works in six different locations from the north to southeast coast and inner China (Figure 1.1) demonstrates important variations that reflect location specific attributes within a global hierarchy of production sites. A total of 53 national-level science and technology industry parks have been established at different locations throughout China (Figure 1.2). Division of the cities between parks featuring predominantly multinational companies and those promoting local innovation emphasizes the two-track nature of development. This study concludes with suggested features of a best practice model suitable for China's circumstances.

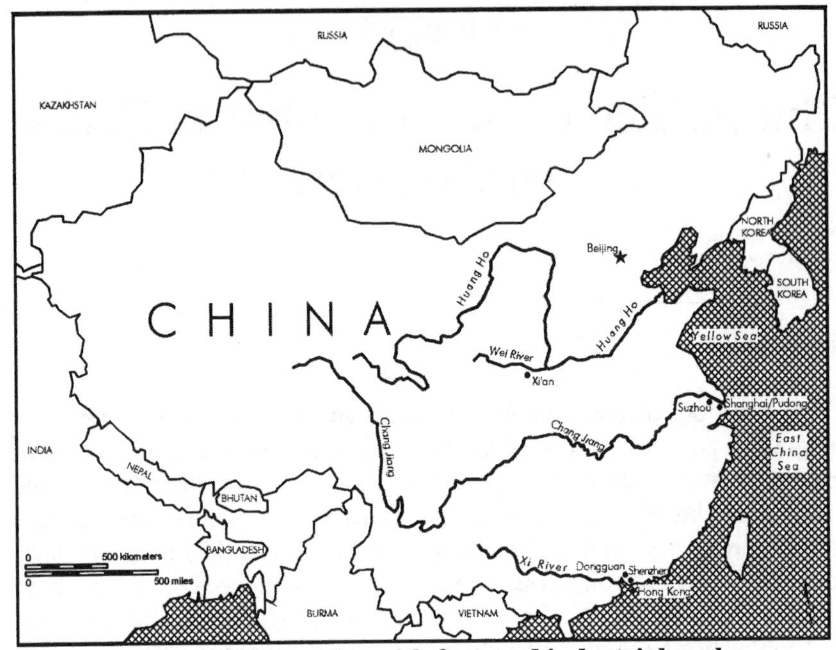

Figure 1.1 Chinese cities with featured industrial parks

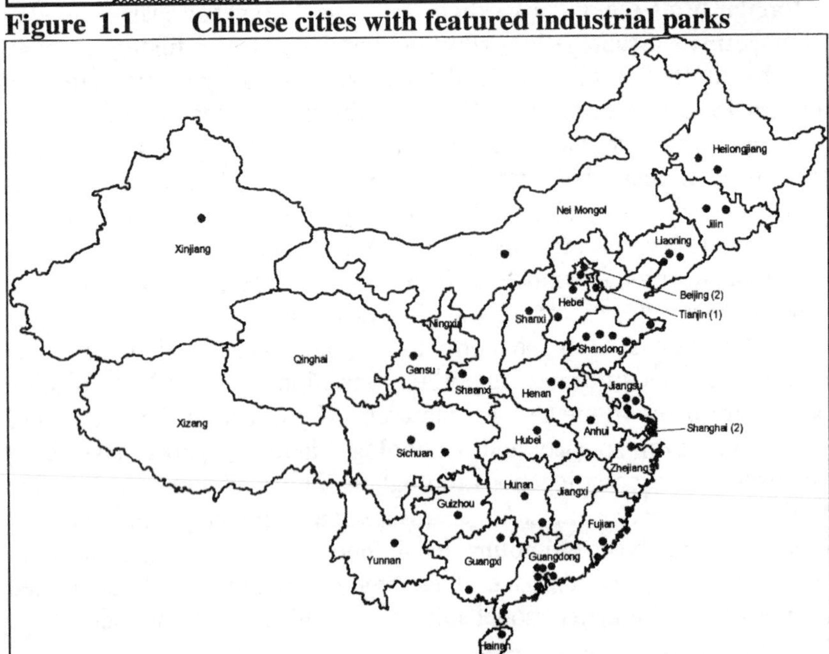

Figure 1.2 Chinese science and technology industrial parks

Repositioning for New Growth: High Tech as Economic High Hope

The first chapter following this overview deals with general theoretical issues within economic geography that frame the investigation. Key questions posed by this examination consider whether the current state of China's science parks falls into previously proposed models, presents a developing world modification, or constitutes a type of its own. Use of High Tech parks as a growth engine reflects a global economy that particularly rewards knowledge-based inputs.

A global hierarchy structures a network of cities wherein regions closest to innovation sites garner the greatest profit, down to regions of routine, low-cost labor who at least are part of a high technology-infused commodity production chain. Local industrial districts are set within geographically expanding global corporate and urban networks. Particular cities serve as major capital accumulation and transaction points. Municipalities and their regions are not immune from the influences of the nation-state, particularly in a country as centrally controlled as China. The success of the so-called Asian Tigers demonstrates that national government policy plays a significant role in shaping the outcome of local development attempts. The role of each participant is thus considered at a variety of geographic scales.

Regions

The economic condition and power of regions rise and decline in succeeding waves. Successful businesses, and the individuals who work in them, consequently are attached less to a place than to a habitat of corporate culture, which reflects characteristics that are both local and transferable. In order to obtain the growth and economic dynamism emanating from international links created by successful businesses, countries and cities harboring such inhabitants compete by replicating a cocoon of familiar internationally interchangeable settings. Economic islands are unsustainable. Highly successful but small scale city states such as Singapore must extend their economy, becoming major sources of assistance to near neighbor China. A network of city-regions, inhabiting but less subordinated to nation-states than in the preceding century, increasingly asserts a new hierarchy of connections in a global production chain, which can be either producer driven or market driven, with power in both cases residing with the more developed country (Gereffi, et al., 1994).

Firms

Corporate mergers occur across country borders, submerging cultures into a new logo identity, while city-regions increasingly scramble to attract lucrative relocated headquarters. Although there is some debate as to the degree of control, consensus prevails as to the relative power of China's central government to decide what areas of the country will receive powerful development incentive priority (Fan, 1997; Lin, 2000). Under current conditions of loosened centralized control, gray areas exist leaving regions to find their own way to economic success. If the outcome achieves the desired ends, the path must have been correct (Cooke, et al., 1998; interviews).

Transnational corporations (TNCs) function to spread an internal corporate culture and organization across geographic space, in a transfer of knowledge enhancing local dynamism. Clustering firms in a district draws on advantages that come in general from being in an urban area (improved infrastructure, larger and more diverse labor base, financial instruments, and information access) and advantages from being in proximity to firms with similar production or distribution affinities. Three criteria, now relaxed since their original promulgation, generally guide Chinese acceptance of foreign direct investments (FDI). Companies seeking to enter China should be technologically advanced, possess the potential to generate foreign currency, and combine both urgent need and the inability of China to produce the product domestically. Recognizing that every investment decision has a spatial component, the goal is for the area's pool of China's highest skilled labor, improved infrastructure and service sector to produce agglomerated economies of scale that will attract foreign direct investment. A 'virtuous cycle' might then occur with labor and capital attracted to a good investment site. Indeed, the value (both contracted and fulfilled) of projects attracting foreign investments has climbed steadily since the rise of market liberalizing advocates in 1976, especially in the 1990s with economically booming times in the West.

Countries investing in China do indeed represent the most developed nations, such as the 'Triad' of the United States, Europe, and Japan. Regional leaders Korea (ROK), Taiwan, and various source-shielding capital havens in the Caribbean (Virgin Islands, Bermuda, Cayman Islands) also represent significant sources of foreign direct investment (Figure 1.3).

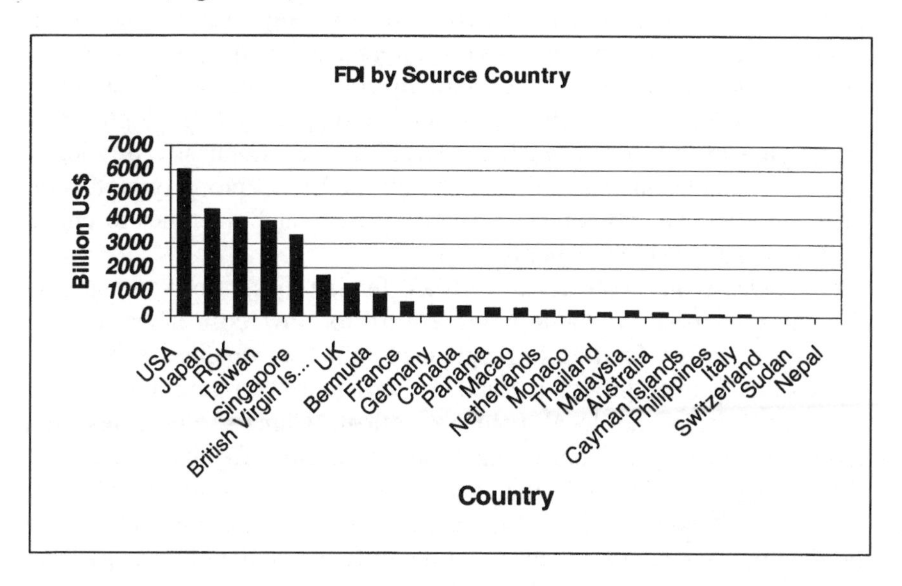

Figure 1.3 Foreign direct investment by source country, 1998

Districts

The term 'technopole' applies to planned developments assisted by local universities, research institutes, land developers, and governments at all scales (Castells and Hall 1994). Technopole enterprises utilize locally created information and scientific labor, though the manufactured commodification of that knowledge may occur elsewhere. Since new ideas generated from research in technopoles draw upon research centers located within or close to parks with companies developing their products (Southern Growth Policies Board, 2002). The land development piece can originate from the private sector, universities, government entities at various scales, or combinations of these. At their core parks are places where research ideas are developed into products for the market. The parks are spatially distinct economic ventures supported by political and educational bodies in recognition of their profound potential for contributing corporate profits to reshape their region.

The ways in which science and technology industry parks (STIP) in emerging economies vary in components and functions from their spatial counterparts in developed countries forms a key focus of this investigation of Chinese parks. In countries with developed economies, places where individuals network pre-existing institutions, from universities to financiers and manufacturers, can be enough over time to create an industrial cluster. In developing countries such as China, much more is needed in the form of government support and development of underlying institutions to support activities in economically privileged spaces. Examples include enforcement of intellectual property protection laws, an effective and solvent financial system, and amenities to entice expatriate native skilled labor to return.

Questions arise as to whether foreign firms keep technopole functions overseas while utilizing China for low cost labor, or use China's scientists for breakthrough ideas but then obtain a substantial share of the profits through partnership arrangements in return for capital needed to keep the local company afloat. China's STIPs test the hypothesis that spatial clustering is an especially effective strategy in developing countries and other regions where barriers to access exist, despite the demonstrated fact that information exchange constitutes a key production element. One measure of the different nature of these parks is to compare the proportional mix of capital from government sources, private domestic pools, multinational companies, and Chinese multinationals. Looking at the types and extent of supplier-labor-market linkages can assess embeddedness. China's technopoles function as a national variant of the model, with distinct characteristics due to China's centrally controlled political economy and early stage of transition development.

Examinations of science parks in various global locations display a wide variety of conclusions as to their usefulness in promoting either business or regional development success. The most negative views see these land set-asides as principally a scheme of real estate developers to enhance tax revenues (Massey, et al., 1992). Other studies, along this continuum, found that park occupants justified the premium rents based on the benefits from spatial agglomeration fostered by the prestige location as well as frequently theorized clustering benefits from enhanced infrastructure to information exchange interactions (Staudt, et al., 1994; Sternberg, 1996).

Studies by the OECD (1987) asserted, and the International Technology Research Institute (Boulton and Kelly, 1999) confirmed, that

> Where extending advanced infrastructures across an entire economy is too costly and complex to accomplish in the short term, science and technology 'parks' have allowed countries to rapidly build state-of-the-art mini-economies within a geographically restricted area...[giving] Asia's high-tech industrial parks an advantage in attracting or keeping global competitors (pp. 60-61).

Networks

Networks knit together the constituent parts of a production chain within and across geographic areas. The nature of these networks falls into two basic schools of thought. One stresses transaction flows, whether of material or information (Storper, 1997; Cooke and Morgan, 1998, Boulton and Kelly, 1999). The other emphasizes the relations of participants who construct, operate and function within networks (Thrift, 1996; Yeung, 2000a). Networks examined in this study are particularly complex, consisting of transnational businesses operating according to procedures set out by their company, within the occasionally clashing cultural norms of their home setting and their host foreign setting. Science and industrial technology park zones combine elements of both worlds by encompassing a variety of interacting geographic scales: the global political-economic macroscale, the nation-province-city mesoscale, and the corporate microscale (Lin, 2000).

Networks for information exchange were objects of persistent inquiry in the interview process. Lack of interfirm connectivity could be a major impediment hampering the blossoming of China's hi-tech parks into something more than zones for extraction of profit for transnational companies. Network factors missing included mediating actors and institutions, rules of engagement, more tech transfer mechanisms, and informal interaction spaces. Other major concerns cited were shortage of venture capital (and capital for investment in general, as well as mechanisms for doing so) and over-extensive bureaucratic restrictions. This section on individual Parks examine to what extent each served some of these needs; it should be noted that certain shortcomings derive from remediable institutional deficiencies during transition from a controlled economy, while others are apparent in some Western districts as well.

Formerly prosperous regions can be regenerated through connection to a global network, particularly as the region draws on high growth, high profit technology-intensive industries utilizing local factors such as labor with suitable cost, sufficient numbers, skills, and/or work ethic attributes (Scott and Storper, 1987). The Rustbelt Midwest's 'Auto Alley' revs new economic engines thanks to Japanese and German mini-steel mills and car manufacturing facilities. Deep South states have turned from textiles to telecommunications. Southern California cities weaned from government defense grants grow biotech upstarts. And across the Pacific, China has overcome a century of civil strife to sprout construction cranes for instant cities from former agricultural fields as in Shenzhen and Pudong, on the outskirts of Hong Kong and Shanghai, respectively. The combination of embedded and created characteristics leads to a politics of place competition between locations and firms illustrated by the case studies in this research, wherein local advantages furnish the competitive features utilized by the wise metropolis and intelligent firm seeking the best location match (Stopford and Strange, 1991).

The contrast in FDI absorption and type can be illustrated by a proportional comparison of the municipality of Shanghai and China as a whole (Figures 1.4, 1.5). In the highly urban setting of that east coast commercial giant, by the end of 1999 the tertiary sector represented 46 per cent of total foreign direct investment, with real estate comprising 54 per cent of that piece. Secondary investments were overwhelmingly in industry, reflecting Shanghai's traditional strength. Primary sector interest was only .2 per cent of the total FDI in the city. For China as a whole, 63 per cent of foreign direct investment was in manufacturing, four per cent in the primary sector, twelve per cent in real estate and six per cent in electronics and telecommunications. Revealingly, only .02 per cent of FDI fell into the category of scientific research. For the purposes of this examination, these figures indicate the interest of foreign investors lies far less in actual innovative R&D using China's intellectual labor but rather in manufacturing supplied by low cost labor.

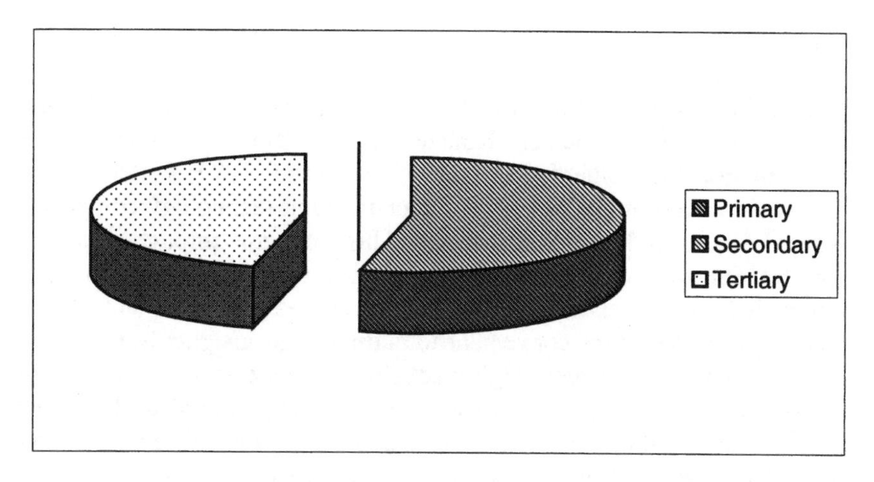

Figure 1.4 Foreign direct investment absorbed by the end of 1999, Shanghai

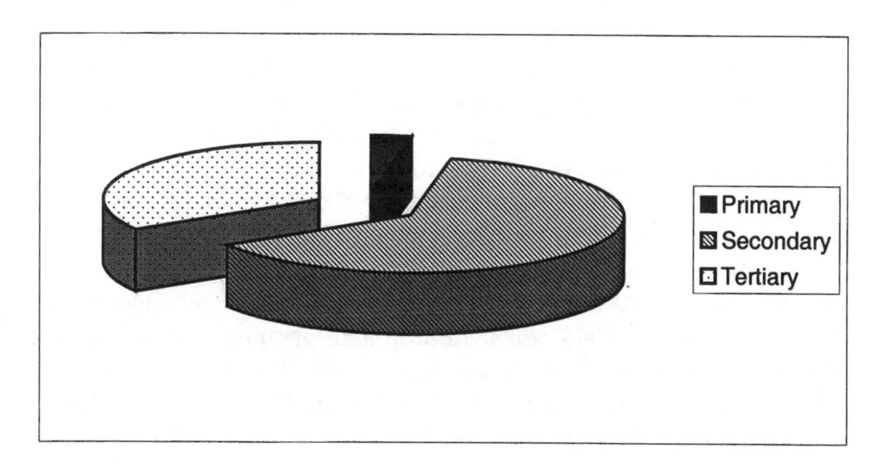

Figure 1.5 Actually used foreign direct investment, China 1997

Globalization

Within the global flow of capital, products, processes, individuals, and ideas, places with particular characteristics are more attractive sites of accumulation than others (Park, 1996; Markusen, 1997; Porter, 2000). The basic underlying competitive factor for high technology companies is rapid access to the best information. This can come through obtaining the most innovative and task-suitable employees, associating with other innovative companies through business, social, or site arrangements, or connecting to locations convenient to cutting edge insights from nearby research centers. The most highly sought economic salvation comes in the lab coat of research and development. Restless capitalism looks for the newest breakthroughs to generate surging market demand and profitable paybacks, with enormous margins that more than cover the costs of past and future product discovery and development. The United States' pre-eminence in technology, and in particular the role played by companies on the West Coast, inspires numerous attempts by other nations to create their own technopoles. Any lack of success to date indicates that essential, identifiable ingredients must be missing. This is less a problem of place, since technological business success is not culture-bound, than a problem of an incomplete understanding of the process. Given the cyclical nature of economic and political change, dynamism is clearly geographically mobile. With microwave communication enabling almost instantaneous transfers bounced off satellite dishes, business is a global proposition.

China's economic relationship with the U.S. greatly assisted that country's meteoric economic development in the past two decades, during which China's international goods trade zoomed from US$28 billion in to US$510 billion in 2001; U.S. trade with China alone grew for US$1 billion in 1978 when 'Reform and Opening' was initiated to US$119.5 billion in 2000. At the turn of the 21st century, the trade deficit with China represented 19 per cent of the U.S.s total trade deficit. While exports to the U.S. constituted over 41 per cent of China's total, the U.S. exported only two per cent of its total global exports to China.

The decade of the 1990s marked a particularly significant change in the nature as well as the volume of the bi-lateral trade relationship between the two countries. Not only did the trade deficit soar from US$11.5 billion to US$83 billion, the proportion of high technology exports shifted from roughly non-existent in 1990 to US$13.3 billion by the year 2001, encompassing over two-thirds of all items on the U.S. 'advanced technology products' (ATP) list. This shift began in 1995, with the ATP trade deficit between the two countries widening from US$1 billion to over US$8 billion by mid-2002. Globalization's influence enhancing the trend for foreign subcontracting enhances China's attraction as an 'export platform' for companies headquartered in more advanced (high cost) countries. In 2000, China was #7 in exports (US$249.3 billion), #8 in imports (US$225.1 billion), and #1 in FDI (US$46.8). Almost all of the ATP trade with China falls into four categories: mechanical equipment including computers (46 per cent), electrical machinery (35 per cent, aircraft and spacecraft (12 per cent), and optical-photographic and measuring equipment (almost 7 per cent) (U.S.-C.S.R.C., 2002). Areas of major U.S. foreign direct investment in China by 2000 are detailed in Table 1.1. Areas of major expansion in China's ATP manufacturing strength include telecommunication equipment and silicon wafers (largely due to cross-Straits re-location by Taiwanese firms). Clearly, China is a major economic player in global trade and investment, meriting further examination of past trends and future potential (Lin, 2000).

Studies set out in this book turn agglomeration theory around. In a free market, companies choose to congregate in order to maximize their utilization of local advantages, thus realizing 'localization economies'. In China's centrally controlled economy, location choices are constrained and highly influenced by government directives encouraging companies to locate in designated areas. This research assesses the outcome of such compelled proximity. Two tracks exist for businesses in Chinese economic development zones: one is for transnational companies (TNCs), while the other is for native Chinese companies. Separate sections treat areas that tend toward each of these types. A model emerges of economic development more suitable to developing countries that usually exhibit a stronger government rather than market directive role.

Table 1.1 U.S. foreign direct investment in China

Sector	Investment (US$billion)	%Total	% Increase (1994-2000)
Electrical equipment Manufacturing	3.2	33.5	1,787
Petroleum	1.8	19.3	106
Financial services	1.1	11.6	179
Machinery manuf.	.93	9.7	NA
Chemical manuf.	.24	2.6	11
Metals manuf.	.18	1.9	76
Food manuf.	.18	1.8	38
NEC manuf.	.77	8.0	252
Wholesale trade	.36	3.8	168
Other	.59	6.2	357
TOTAL	9.57	100	275

Source: U.S.-CSRC, 2002, based on US Bureau of Economic Analysis

Chinese Science and Technology Industry Park Policy

Moving from macro scale issues of theory to the particulars of China's setting, the third chapter discusses how Deng Xiaoping's economic reforms opening China since 1978 to more capitalist ways signaled the beginning of a major shift permitting closer integration of the domestic economy with the global economy. Subsequent development plans targeted particular sectors for investment and attraction of foreign companies. In 1988, the launching of the national 'Torch Program' provided carrots (incentives) and sticks (regulations) for luring foreign investments to 54 'economic and technology zones' and 27 science and industrial parks concentrated largely in coastal China. These sites represented, in the judgment of the central government, the areas most likely to be able to attract, absorb, and thrive from infusions of foreign capital in the form of corporations and managers. By 1997, foreign funded enterprises represented significant investments in these zones throughout China, particularly along the eastern coast. The large concentrations of companies in Beijing, Shanghai, and Shenzhen-Guangdong development areas were especially successful in securing High Techincome, though 'technopoles' operate in large cities with major research institutes throughout China (Wang, et al., 1998).

Comparison Models: Multinational Development Zones

Various new types of districts dot the economic landscape of developing countries such as China, posing variants on technology park and corporate configurations due to their transnational nature bridging developed and transitioning economies (Park, 1996). This study looks at several zones in geographically diverse regions of China, noting differences due to the type of activity contained and the type of company profiled. The comparative approach combines a manufacturing zone framework within a regional analysis (Johnston, 1999), arguing that similarities within zone type can still be usefully contrasted by regionally influenced factors.

The first set of three multinational development zones considers the role of transnational companies as economic development growth engines for China's leapfrog into the 21st century network of global capitalism. China's post-1978 modernizers see this as the fastest way to gain access to transferable Western technology, while financing the development of local companies through foreign trade profit. Since the initial proclamation of an 'Opening and Reform' movement in 1978, the number and value of projects involving foreign investment has increased at a steeply accelerating rate (Table 1.2). The slowdown from 1989-91 reflects both the general global recession and the chilling effect of the Tiananmen suppression on foreign investment. Consequently, the huge increases of the next two years demonstrate not only recovery in the American economy, but the sense that China was again secure for investment and eager to catch up quickly with delayed opportunities.

China's economy also benefits by an increase in jobs in the foreign-invested sector while China's bloated state-owned enterprises (SOE) dismiss workers as part of an unprecedented restructuring. The types of economic investments made by foreign companies reflect a specialization in areas most affected by modernization infrastructure needs, and most neglected in the preceding decades: real estate services, electric, gas, and water infrastructure, electronics and telecommunication facilities. Foreign companies often come to China to gain a foothold in what is perceived as a potentially vast 1.2 billion market, presently providing a large, inexpensive labor pool for simple assembly operations. Chinese workers come to foreign enterprises for the better pay and generally better working conditions, which varies widely depending on the country of ownership.

The degree to which these two expectations coincide depends on the speed with which China develops a middle class with expendable income.

Table 1.2 Foreign capital actually used

Year	Total used (US$ 100 million)	%Change
1979-83	18.02	
1984	12.58	(33)
1985	16.61	31.0
1986	18.74	12.8
1987	23.14	47.0
1988	31.94	38.0
1989	33.92	6.0
1990	34.87	2.8
1991	43.66	25.0
1992	110.07	152.0
1993	275.15	150.0
1994	337.67	22.7
1995	375.21	11
1996	417.26	11
1997	452.57	8
1998	454.63	0.4
1999	403.19	-11.3

Source: Adapted from China Statistical Yearbook, 1998

High technology industrial zones are key places for the growth of a middle class, particularly one featuring new skills in management, science, and entrepreneurship. These zones are geographically concentrated heavily in the eastern section of the country, as demonstrated in Table 1.3. As the computer technology center, Beijing generates by far the most tech-based revenue, while Shanghai demonstrates its strength as an export center. Xi'an's technology, as shown below, flows mainly into domestic ventures rather than export.

Aggregating the four major economic regions in China provides a picture of their relative technology strength (Figures 1.6, 1.7). The Yellow River delta, including Beijing, Tianjin, Hebei, Liaoning, and Shandong, contains 12 State-level SHIPs.

The Pearl River delta includes the provinces of Guangdong, Guangxi, Fujian and Hainan, and contains 11 SHIPs. Shanghai, Jiangsu, Zhejiang and Anhui are in the Yangtze River delta, with seven SHIPs. The Western region contains Shaanxi, Shanxi, Inner Mongolia, Sichuan, Chongqing, Guizhou, Yunnan, Gansu, and Xinjiang, with a total of 12 SHIPs.

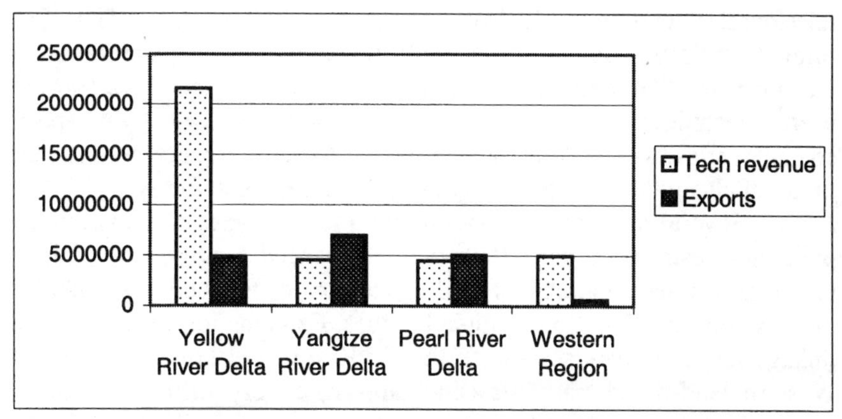

Figure 1.6 Regional distribution of technology-related revenue (1,000 RMB) and exports (USD), 2000

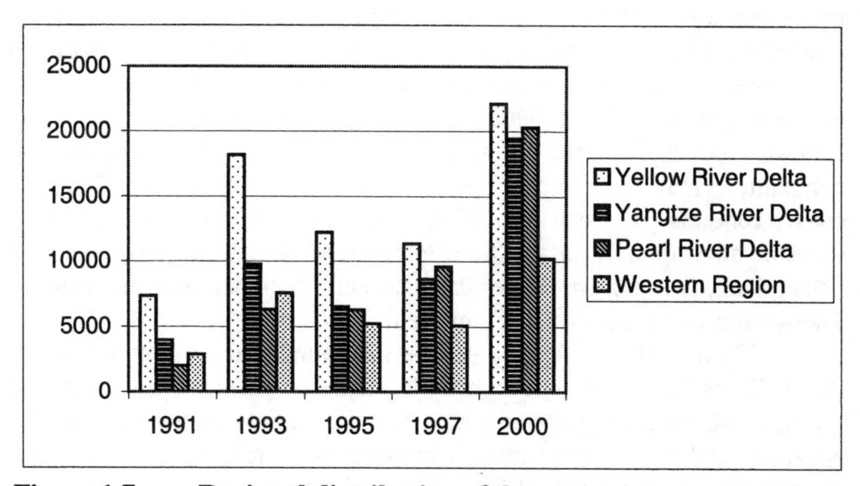

Figure 1.7 Regional distribution of domestic patents granted

Some of the regional differences in patents flow from discrepancies in the number of inventions, the general awareness of patents, and the prevalent size of companies (small ones, as in Guangdong, are more likely to seek patent protection).

Both parties operate warily against the backdrop of a long history of engagements, from silk for 1st century Caesars in the Han Dynasty, to Marco Polo and Mateo Ricci, marauding merchants and missionaries, and multiple invasions in the 20th century. The 'Self-Strengthening Movement' in the 1890s featured bureaucrats embroiled in a controversy that continues in the present day: how can China learn the useful techniques of the West, while keeping her own special characteristics? China remains determined to set the pace (gradual) and limits of change in any sphere, opting in all cases for stability.

Several places serve to represent different types of economic and political designs, given the richly nuanced nature of 'place'. Corporations are creatures as well as creators of their context; Western models currently function as global guides. Examination of the dynamic linkage model moves across the Pacific Rim to see how an essentially Western land-based park set-aside fares in a very different political-economic environment. China represents the common situation of many countries, transitioning from a tightly controlled economy and centralized government to seek rapid development as a global player. For more than a decade this populous country of 1.2 billion, physically the size of the United States, has exerted a magnetic pull on transnational companies seeking a foothold in a potentially huge market of hard-working, low wage labor. This research examines sample cities in north, central, south, and interior China provide insights into how multinational companies create - or vacate - environments to sustain them. The east coast contains an overwhelming concentration of foreign direct investment, primarily in the three regions of the Pearl River delta in Guangdong, the Yangtze River delta around Shanghai, and the Beijing-Tianjin axis in the northeast (Figure 1.8).

Each district reflects a unique institutional and cultural context, as well as generally recognizable features. Local and national governments seek to energize a network of suppliers, deliverers, service providers and buyers from foreign corporations – a spark that economic planners hope to fan into Mao's proverbial forest fire. Western High Techfirms serve as the focus of this inquiry into science park suitability for serving as economic engines, given the reputation of Western technology as relatively available for sharing and compatible with China's ability to digest at the current stage of sophistication. The generation of synergy is the most difficult step, especially within China's framework of information control and companies from multiple foreign countries. The emerging economic strategy of cooperative competition helps to build critical bridges, with companies looking for a competitive edge breakthrough as a base to build critical mass dominance in some area. Local advantages of a science park could supply this edge, particularly in a foreign setting where transnational companies need intercessors and interpreters.

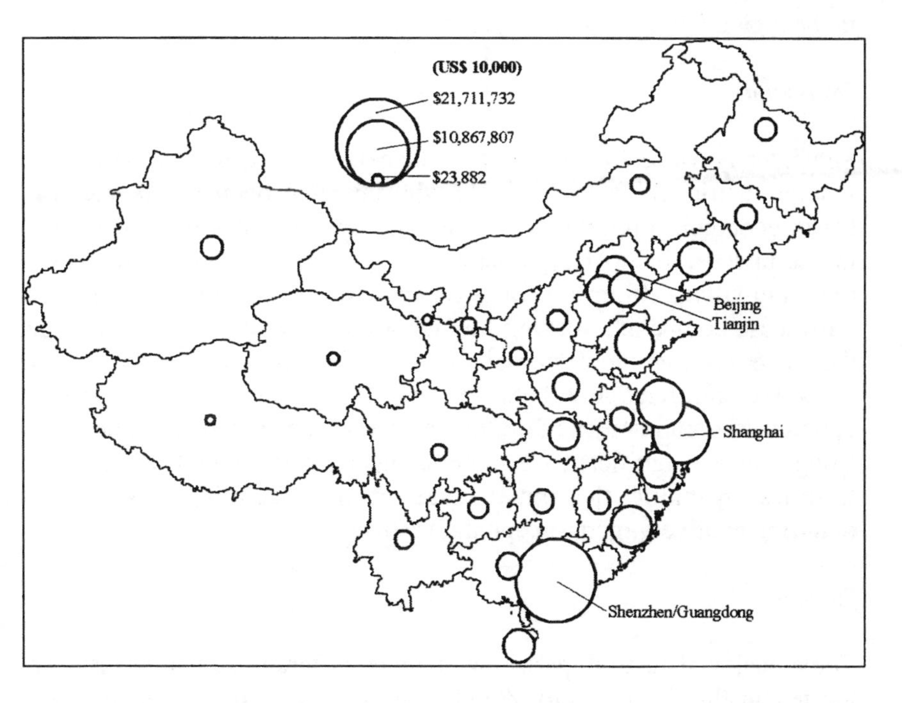

Figure 1.8 Location and amount of foreign direct investment

Comparison Models: Multinational Development Zones

Shenzhen

Outside the large south China city of Guangdong, the Shenzhen Science and Industry Park was founded in 1985 both to balance developments in the north and to tap into foreign capital and developments from nearby Hong Kong, fueling the growth of modern China's first new city. The city government, the County Investment Corporation, and the Chinese Academy of Sciences administer the park jointly. Shenzhen is especially successful in attracting overseas Chinese investment, and low-cost labor ventures of other foreign companies. Many studies have demonstrated that Hong Kong ties principally flow from personal and family networks in the region (Wu, 1999; Yeung, 2000B).

Dongguan

Strategically located to the west of Shenzhen, Dongguan is the historic sending district for many Chinese who migrated overseas in search of better economic opportunities. With the opening of China, they now invest in local industries, assisting the growth of a highly successful township-village enterprise. Dongguan also hosts a cluster of Korean and Taiwanese companies seeking low labor cost advantages by employing the numerous migrants (especially unmarried youths) from inner Chinese areas seeking relatively better paying jobs than those available in agricultural communities of origin, or slowly industrializing inland cities. Dongguan is a significant source for computer parts assembly and textile factories, spanning the initial stage of manufacturing development requiring relatively modest capital investment.

Suzhou

Three major industrial parks, with quite different constituencies, are located in this historic city 80km and two hours from Shanghai. The China-Singapore Suzhou Industrial Park (CSSIP) sits on the western edge of the city, the first park structured as the result of government-to-government agreements between Singapore and the P.R.C. Over 90 per cent of its occupants are foreign companies. The Suzhou Industrial Park, more closely aligned with the provincial government of Jiangsu, grew more rapidly in the previous several years. Kunshan Industrial Park, like

SIP on the eastern side of Suzhou closer to Shanghai, contains many companies aligned with overseas Chinese interests such as those headquartered in or owned by Taiwanese.

Comparison Model: Multinational Learning Zone

Shanghai

Shanghai represents the most likely site for success of the endeavor to leapfrog China's development via High Tech transfers from foreign companies. Four out of thirteen economic development technology parks, designed with varying success to capture High Techforeign companies, are investigated in greater detail. Minhang is a 'satellite city' suburb of Shanghai, created to lessen congestion in the central city. Schering-Plough's directed location in Minhang triggered creation of the first area ETDZ in 1985. Caohejing Hi Tech Park followed three years later, as then-Shanghai Mayor (now China's Premier) Jiang Ze-min's design for an in-town Sino-Silicon Valley housing specifically hi-tech companies. The latest Hi Tech park, designed to attract life science industry enterprises, grows gradually on the southern edge of developed Pudong. Connected to Shanghai by three bridges, a subway, and a new international airport, Pudong's mission is to grow Shanghai to its eastern-most extension, housing modern financial, industrial, and expatriate facilities.

Comparison Models: Local Innovation Learning Zones

The next set of three development areas distinguishes those more reliant on domestically generated technology, as well as interactions with foreign companies. Beijing's Zhongguancun district, characterized by university-affiliated entrepreneurs, utilizes high technology industries along the lines envisioned by government agencies hoping to create districts configured to encourage the growth of indigenous enterprises (Wang and Wang, 1998). Development districts in Shanghai, a 'globalizing' rather than fully integrated global nodal region (Amin and Thrift, 1992; Yeoh, 1999), can be classified along a continuum of both of these tendencies. The China-Singapore Suzhou Industrial Park typifies the former and Caohejing High-Tech Park the latter. The location of China's premier educational institutions in Beijing lead to the relatively high innovation input of native High Tech companies there, while Shanghai's attraction of foreign companies predisposes that area to foreign technology transfer attempts. Shenzhen's design across the border from Hong Kong not only attracts a hugely disproportionate amount of FDI to Guangdong from Hong Kong, but also impacts the relatively efficient one-stop, red tape cutting approach of Shenzhen's high technology park (Wu, 1999). Xi'an's relative isolation, in contrast, means it must draw more heavily on local college resources and its historically close ties to China's military-industrial complex.

Beijing

The first technology transfer-based company was founded in Beijing, based on a 1980 discovery by a professor at the Chinese Academy of Sciences. Five years later a hundred electronics related enterprises lined 'Electric Street' in the northwest city quadrant closest to several major universities. Beijing's 'Experimental Zone for the Development of New Technology Industries' includes the Zhongguancun 'Science City', 138 research institutes, and 55 colleges in a 100 sq. kilometer area particularly noted for computer software development (Wang, 1999).

Shenzhen II

Shenzhen maintains strong ties with educational institutions in Beijing, attracting research-intensive companies such as pharmaceutical, IT, and engineering firms to Shenzhen Hi-tech Park that draw heavily on technology transfers from northern universities. A separate section within SHIP features a University Park to put related facilities in close proximity. Assisting Shenzhen also serves to balance resurgent (and North-South rival) Shanghai in the central Changjiang delta, inserting an innovation-promoting element to offset low labor-intensive industries in outlying Dongguan.

Xi'an

The Xi'an New Technology Industrial Development Zone is one of seven districts in this former Silk Road home base. Xian also served as a military-industrial center interior base in the pre-Reform era, continuing a historic military tradition demonstrated by the famous life-size pottery militia of the Yellow Emperor from the Great Wall building era. The corporate presence ranges from Coca Cola and Mitsubishi to an International Entertainment Building. Typically, the local branch of a national engineering school (Jiaotong) recently opened its own nearby technology park to nurture local collegiate technology ventures through their commercial market debut. The success of west-central China's 'Silicon Valley' remains to be seen.

Modeling Best Practice

Combining numerical with interview and survey measures provides a picture with lessons, cautions, and promise for all involved. Analysis of the business environment in each setting and city-region is based on extensive interviews in each site. Subjects were targeted based on several criteria. Local development entities – in some cases land, political and/or industry-based – were contacted for lists of relevant companies in the area, as well as recommendations as to key players. Each subsequent interview included with a request for other names to contact, building an understanding of area dynamics that is more thorough than previous statistical-based studies. China hopes to use Western models of geographically configured cluster development to build a more prosperous future, encouraging modern enterprises based on brains as well as low wage labor pools and a potentially huge domestic market for domestically made goods and services. Western companies foresee a large foothold on the Asian market, as China ranks first with more than half of all investors choosing sites for both investment and production.

Sites vary greatly in their productivity, however (Table 1.3). While enterprises in high technology zones in the Eastern region of China clearly predominate, with more than half to 3/4ths of the country's total in each category, the individual character of each city's output varies widely. While Beijing has a huge lead in numbers of companies, workers, and tech-related revenue, Shanghai predominates in exports. Likewise, Xi'an eclipses Shenzhen in the first three categories, but Shenzhen towers over Xi'an in exports – clearly due to their vastly different locations, with the former next to Hong Kong and the latter deep in inner China. Suzhou's export orientation, almost equal to Shanghai, reflects its location close to that major port metropolis.

Table 1.3 Development of high technology industrial zones

Area	Company	Workers	Tech	Exports (US$1,000)
Beijing	6,181	291,473	16,308,165	1,816,922
Shanghai	434	90,563	1,840,269	2,327,534
Suzhou	236	69,038	16,279	2,249,153
Shenzhen	122	63,606	77,122	2,085,795
Xi'an	1,590	97,738	1,202,651	104,079

Source: 'China Statistical Yearbook (CSY) on Science and Technology (S&T)' (2001)

A more insightful comparison of the featured STIPs comes from using the data shown above to figure their relative technology revenue productivity per worker, and per worker export value produced. This evens the field for the larger and smaller cities, producing a clearer picture of which are more export-oriented (Table 1.4).

Table 1.4 Worker technology productivity, 2000

	Average # of Tech Workers Per Company	Tech Revenue Produced per Worker	Exports (US$) Produced per Worker
Beijing	47.16	55.95	6.23
Shanghai	208.7	20.32	25.70
Suzhou	292.5	.24	32.58
Shenzhen	521	1.2	32.79
Xi'an	61.5	12.3	1.06

Source: Adapted from CSY on S&T (2001)

Successful regions for High Tech companies are geographically configured to promote the following five factors crucial to a dynamically integrated model:

- Networks support rapid exchange of information both formally through structured events and informally in non-business oriented settings, from shared meal facilities to community and family events.
- Familiar living conditions create a recognizable package of amenities: international schools, recreation outlets, arts facilities, large homes.
- Basic physical infrastructure is dependably delivered: electricity, telephone and fax lines, Internet, surface transportation, water, etc.

- Government institutions at all levels support the growth of these businesses, who also justify their support by demonstrating their contributions to the general welfare of the region.
- Advocates keep the concerns of members before those who can address their agenda, actively promoting connections.

An industrial park-based high technology learning district best practice model encourages the creation of dynamic clusters of companies whose success can then ripple through the economy at multiple scales. Major examples are utilizable and in place in actively growing city-regions throughout a global economically linked network inhabited both by transnational entities and individuals. The following chapters examine to what extent these conditions are present in contemporary China, and where they are currently being constructed. Each chapter concludes with a summary of the factors supporting the success of STIPs in that location. The final chapter discusses the macro-level national framework as it impacts STIP development, focusing on market institutions and policy initiatives that would assist a Best Practice model. The repercussions of China's entry into the global economy as a location with a large market fuelled by low cost production labor and high skill capacity has a huge impact for East Asia and the global economy.

References

Amin, A. and Thrift, N. (1992), 'Neo-Marshallian Nodes in Global Networks', *International Journal of Urban and Regional Research*, vol. 16, pp. 571-87.

Boulton, W. and Kelly, M. (1999), 'Information Technologies in the Development Strategies of Asia', *Paper for the International Technology Research Institute*.

Cooke, P. and Morgan, K. (1998), *The Associational Economy: Firms, Regions, and Innovation*, Oxford University Press, Oxford.

Cooke, P., Uranga, M. and Etxebarria, G. (1998), 'Regional Systems of Innovation: An Evolutionary Perspective', *Environment and Planning A*, vol. 30, pp. 1563-84.

Fan, C. (1997) Uneven Development and Beyond: Regional Development Theory in Post-Mao China, *International Journal of Urban and Regional Research*, vol. 21, pp. 620-39.

Gereffi, G., Korzeniewicz, M. and Korzeniewicz, R. (1994), 'Introduction: Global Commodity Chains', *Commodity Chains and Global Capitalism*, Greenwood Press, Westport, CN.

Johnston, M. (1999), 'Beyond Regional Analysis: Manufacturing Zones, Urban Employ and Spatial Inequality in China', *The China Quarterly*, vol. 157, pp. 1-17.

Lin, G. (2000), 'State, Capital, and Space in China in an Age of Volatile Globalization', *Environment and Planning A*, vol. 32, pp. 455-71.

Markusen, A. (1997), 'Sticky Places in Slippery Space: A Typology of Industrial Districts', *Economic Geography*, pp. 293-310.

Massey, D., Quintas, P. and Wield, D. (1992), *High Tech Fantasies: Science Parks in Society, Science and Space*, Routledge, London.

National Bureau of Statistics, Ministry of Science and Technology, ed. (2001), *China Statistical Yearbook on Science and Technology*, China Statistics Press, Beijing.

Organization for Economic Co-Operation and Development (1987), *Science Parks and Technology Complexes in Relation to Regional Development*, OECD, Paris.

Park, S. O. (1996), Networks and Embeddedness in the Dynamic Types of New Industrial Districts, *Progress in Human Geography*, vol. 20, pp. 476-93.

Porter, M. (2000), Location, Competition, and Economic Development: Local Clusters in a Global Economy, *Economic Development Quarterly*, vol. 14, pp. 15-34.

Scott, A.J. and Storper, M. (1987), High Technology Industry and Regional Development: A Theoretical Critique and Reconstruction, *International Social Science Journal*, vol. 112, pp. 215-232.

Southern Growth Policies Board (2002) *Innovation U: New University Roles in a Knowledge Economy.*

Staudt, E., Bock, J. and Muhlemeyer, P. (1994), Technology Centers and Science Parks: Agents or Competence Centers for Small Businesses? *International Journal of Technology Management* vol. 9, pp. 196-212.

Sternberg, R. (1996), 'Regional Growth Theories and High-Tech Regions', *International Journal of Urban and Regional Research*, vol. 20, pp. 518-38.

Stopford, J. and Stopford, S. (1991), *Rival States, Rival Firms: Competition For World Market Shares*, Cambridge University Press.

Storper, M. (1997), *The Regional World: Territorial Development in a Global Economy*, The Guilford Press, New York.

United States-China Security Review Commission (July 2002), *Report to Congress: The National Security Implications of the Economic Relationship Between the United States and China*, Government Press, Washington, D.C.

Wang, J.C. (1999), 'In Search of Innovativeness: The Case of Zhongguancun', in Malecki, E. and Oinas, P. (eds.), *Making Connections: Technological Learning and Regional Economic Change*, Ashgate, Aldershot, UK.

Wang, J.C. and Wang, J.X. (1998), 'An Analysis of New-Tech Agglomeration in Beijing: A New Industrial District in the Making?', *Environment and Planning A*, vol. 30, pp. 681-701.

Wang, S., Woo, Y. and Li, Y. (1998), *Development of Technopoles in China*. Asia Pacific Viewpoint vol. 39, pp. 281-301.

Wu, W. (1999), *Pioneering Economic Reform in China's Special Economic Zones*. Ashgate, Aldershot, UK.

Yeoh, B. (1999), Global/Globalizing Cities. *Progress in Human Geography* vol.23, pp. 607-16.

Yeung, H. (2000a), Local Politics and Foreign Ventures in China's Transitional Economy: The Political Economy of Singaporean Investments in China, *Political Geography*, vol. 19, pp. 809-40.

Yeung, H. (2000b), Embedding Foreign Affiliates in Transnational Business Networks: The Case of Hong Kong Firms in Southeast Asia, *Environment and Planning A*, vol. 32, pp. 201-22.

Chapter 2

New Industrial Districts: Theories and Models

Theoretical Propositions

Interest in 'what firms do, where they do it, why they do it, why they are allowed to do it, and how they organize the doing of it across different geographic scales' (Henderson, et al., 2002, p.5) forms a research focus that is cross-disciplinary within academia and is shared with the business community most directly involved. Geographers focus particularly on multi-scalar components of the territoriality of production, examining elements of the process that are rooted in the context of a particular place (micro-scale firm level) as well as meso-level metropolitan, regional, nation-state, and transnational operational factors. What Castells (1996) proposed as a new 'space of flows' in contrast to its preceding 'space of place' is reconceptualized in this study, based on detailed empirical observations, as 'flows through place', where exchanges (information, networks) occur that are constricted or accelerated by place-based contingencies. The selection of industrial parks as the space for examination of economic activity reflects the interest in comparing operations of flows across different scales of interactions, in places that vary geographically. They do, however, share the common intent to maximize flows (the key competitive ingredient for an innovation promoting milieu) – whether of information, labor, materials, and/or financial capital. The actual type of companies and processes involved in this milieu can vary widely, but the outcome is produced by new innovative behaviors (Hall, 1990).

A variety of theoretical descriptions and explanations have been advanced to capture the location dynamics of the production network and the observed clustering of related industries. Harvard Business School's

Michael Porter (1990, 2000) contributed observations on industrial clusters and their related firm connections along a 'value chain' across a particular terrain. French economists and geographer Michael Storper (1997) amplified this notion to speak of a '*filiere*' encompassing the activities and relations of both humans and commodities in the production and distribution process.

More explicit recognition of the increasing importance of the global scale in production networks led to theories concerning 'Global Production Networks', 'Global Commodity Chains' (Gereffi, et al., 1994), and the more recently proposed 'Global Production Network' (Henderson, et al., 2002). The latter emphasizes five elements that are particularly critical for understanding the interplay of factors shaping firm performance and locational advantage in Chinese science and technology industrial parks (STIPs), the subject of this research.

- First, the priorities of economic actors (e.g., firms, governments, managers of various service entities, universities and related research personnel) often differ, as do their powers and interpersonal relations, and should be taken into account.

- Second, input-output factors vary by location within any given production and distribution (P&D) chain, often constituting key factors in the location decision as to where profit can best be extracted at any particular point and time.

- Third, place-specific contexts must be considered as part of the territoriality of production.

- Fourth, production chains vary depending on whether they are 'producer-driven' or 'buyer-driven'. This is an important consideration in a country the size of China, which supports both an export setting and a huge, presently only fractionally realized, potential market.

- Finally, areas of intellectual property importance, with R&D significance such as in the high technology sector, deserve special consideration. The relevance of this theoretical framework for the current research stems from the background in Asian developmental economies of some of its founders.

Major Themes

Several theoretical strands converge and complement each other concerning the contribution to economic development of investments made to support industrial clusters based on science and technology. The impetus given to local economic growth by locally developed innovations is established in work done by Griliches (1979) and updated by Romer (1990). The economic equivalent of 'success breeds success' when profitable companies attract others to the same location is illustrated by (among many others) Glaeser (1998) and Porter (2000). Work done by Vernon (1966) systematized the importance of fit in the product cycle between locational attributes and the type of product (innovative, mass, declining demand), as well as its extension to apply to global production sites. As a large developing world country, China seeks to maximize its limited investment capital by investing in science and technology ventures in carefully selected regions, further localized in designated science and technology parks. The issues discussed in this book revolve around five central themes:

1. *Place and space* relationships occur at various spatial levels. In this research, for example, regions are treated at the Subnational level, e.g. Pearl River and Changjiang delta areas, Inner China to their west, or Western China as for Xi'an.

2. *Firm* related debates depend on whether the firm is treated as a particular type of organization (e.g., headquarter, branch, transnational corporation) or as a reflection of the individuals in charge.

3. *Firms in places* consider consequences of location within, in this case, a particular industrial park in a particular city.

4. *Network* types and functions constitute the fourth area of theoretical importance. This research examines the interaction of actors, to what extent they support learning, and their locus of identification, e.g., with a place (such as a home village, province, or nation), or on a human level as through kinship, language, transactional, or personal connections. Network participants include place-based interests (nation, urban, provincial, or industrial park scale), economic interests (owners, managers, or workers in firms), and concepts

involved with human elements, such as the 'new international division of labor' (Piore and Sabel, 1984).

5. *Globalization* issues impact networks of places, firms, industries, and humans. This topic continues to generate a great volume of both theoretical and case study examinations due to its current importance and relatively recent recognition as a significant economic-geographic element that touch every place.

Place

To paraphrase geographer Ron Martin (1999, pp. 15-16), the effect of globalization may 'have eliminated *space*... but it has by no means undermined the significance of ... *place.*' Places function as containers of embedded practices, to which entities that would remain there must seek in some sense to adapt in order to survive. Given the continuing discussion of whether the relative strength of global economic business has rendered the nation-state or lower level political units obsolete, it remains clear that in China the nation state's perceived interests prevail.

No natural scale exists for examining economic interactions; the functional level most impacting the particular processes under examination, at the scale of interest, varies by the case. Areas as spatially confined as industrial park technology districts 'have come to be seen as key territorial units in the process of globalization', while their embedded local (urban) characteristics drive 'the economic prospects of a place' (Sheppard, 2001, p.7). The very fixity of such historically conditioned characteristics provide the predictability sought by investors within the global 'space of flows' of possible locations (Castells, 1996). Relative location also matters, as China's current economic location on the near periphery of the core economies indicates it has come of age to be interesting to more mature capital seeking an investment location to maintain shifting advantage, thus attracting international courtship.

Firms

As the characteristics of places are unique and important, so do firms retain inherent headquarter place characteristics, despite the debate engendered by claims to the contrary (Dicken, 1998; Yeung, 1998). Asia furnishes a particularly clear backdrop to display this assertion, where contrasts between Overseas Chinese firms, techniques, individuals, and networks contrast with patterns prevalent on the China mainland, as well as practices common to non-Chinese Asian and non-Asian firms. The case studies set out in the following chapters add more empirical evidence to this frequently illustrated theoretical assertion that firms reflect the cultural practices prevalent in their headquarters.

Theoretical approaches to studying firms classify them in a variety of different ways (Taylor and Asheim, 2001). The *neoclassical* approach assumes that firms pursue their basic 'production function' in a rational and efficient manner with total knowledge of the factors involved. The *transaction cost* school of Coase (1937) and Williamson (1975) stressed the quantifiable cost-benefit calculus that they saw determining economic activities. The *behavioralist* school acknowledges the 'bounded rationality' of firms who are guided by humans acting on the limited knowledge available to them, and asserting their self-interest at key points such as the 'location decision' of where the firm should operate. *Structuralism* sees firms as confined with the predictable coils of larger issues such as capital and class conflicts.

More recent theories integrate socially constructed aspects into the economic functions of firms. The *institutionalist* school of thought cautions that firms are bound by rules and routine behaviors. Socially constructed ties of 'reciprocity and interdependence' (Taylor and Asheim, p.316) control firms in the view of *network* analysts – a viewpoint that seems particularly suitable to the Chinese framework that prioritizes the importance of *guanxi* (connections) relationships. These consist of a mix of pre-existing kinship and/or origin location similarities, and constructed ties of reciprocal obligations. The most innovation-prone firm is one open to *learning* in a learning-supportive area, an aspect that suits much place-based research by geographers studying innovative firms in particular regions, from Silicon Valley, California to the 'Third Italy' of northern Emilio Romagna. Interest in innovative firm organization extends to examining the skill level and organization of the labor force, from fully integrated, top-down vertical Fordist forms to horizontal, sub-contracted and de-skilled labor. The *competency*, or resource-based, school introduces the element of needed resources as a corollary to learning. If a firm, place or individual doesn't have what it takes (oil, coal, amenities, intelligence, connections), then they won't get it (the job done). The postmodernist *discourse* viewpoint accentuates the role of negotiating from positions of unequal power in order to construct meaning. Situations sometimes cause actors to behave in ways less beneficial to the firm as a whole than to clique advancement, based on agreed upon priorities, values, and meanings. Eventually, the firm becomes a temporary *coalition* of interests who adhere to advance their interests.

Place-less theories fail to incorporate the crucial element of geographic influence: where a firm *is* necessarily influences what it does and how it operates, as illustrated in the following case studies of increasingly capitalist China, or market oriented socialism 'with Chinese characteristics'. The key characteristic is the restriction of change to the economic sphere, with the structure of political power held constant. The embedded cultural nature of firm characteristics complicates the process of technological transfer and learning, but not to an insurmountable extent. Since culture is itself learned behavior, a different set of practices is amenable to conscious acquisition. One technique frequently used by transnational corporations (TNC) for integrating their overseas operations consists of intertwining all related entities in a 'relational network' (Dicken, 2000). This speeds the learning process, and can be accomplished by stationing key individuals at headquarters and various other rotations of overseas locations. Place and geography do indeed still impact firms and the economic outcomes of their endeavors.

Firms in Places: New Industrial Districts

The conceptualization of the firm utilized in this book's examination focuses on the socio-economic aspect of firms (Thrift and Olds, 1996) as actors situated in a particular place, but also recognized and self-aware bearers of habits from their headquarters. Such characteristics can include elements that are national – in recognition that firms from Japan, the U.S., and China, for example, do behave differently – corporate, or individual, but reflect a linked cultural component. Places transform the way that firms function in them – but firms also affect the way that places reconstitute themselves to attract particular sorts of firms, creating a nexus of intersecting best interests (Dicken and Malmberg, 2001). This third sub-section deals with the linked aspect of firm and territory, especially in the transnational industrial park form examined in this study.

Modifying a suggestion by Park and Markusen (1995, p.83), this book proposes defining new industrial districts as '*spatially delimited areas of trade-oriented activity which have a distinctive economic specialization based on local factors of production*'. These bounded areas are not only locations where understandings of how business can be conducted are joined, but also places for magnifying the possibilities of new interactions and learning (Massey, 1994). Industrial place theory is largely based on experiences derived from companies in the developed 'Western' world. Increasingly, new studies seek to extend existing theory by reflecting the experiences of capitalist companies doing business in more controlled, developing world environments. Unlike conditions in their headquarter location, the host environment features a stronger role for state planning, an abundance of non-local labor due to migration, and the numerous challenges posed by being a foreign entity.

Traditional Marshallian districts reflect conditions prevalent when they were first described, in late nineteenth century London (Marshall, 1890). Locally tied elements are very strong, including worker loyalty, financing, the scale of inter-firm trade, and the locus of major decision-making. A century later, globalization factors led to the proposal of 'Neo-Marshallian nodes' (Amin and Thrift, 1992). These sites were global centers for transnational companies, seen as possessing key identifiable locational assets (e.g., large pools of low cost and highly skilled labor, a stable political economy, quality of life amenities) that attracted companies seeking other embedded unique advantages as well. The variation seen in the 'Third Italy', the subject of many studies a century later, also featured small firms with strong local ties. A high degree of cooperation among district firms, local associations, and government reflected kinship cooperation-competition enhancing ties. Traditional hub-and-spoke districts are also place-specific for advanced economies with a locally dominant large firm and a cooperative local government. The final type is a newer variety, the 'pioneering high technology district' (Park, 1996) where companies congregate to enhance their success at producing new innovations.

China's specific setting leads to a proposal modifying the preceding new industrial district forms (Figure 2.1). This typology will be fleshed out by case studies in the following chapters. The first of three STIP models is a Transnational Satellite Platform. On the outskirts of a town, the largely foreign firm occupants operate with no significant local university connections. Local suppliers, found by the park authorities or companies, locate on the periphery or close by.

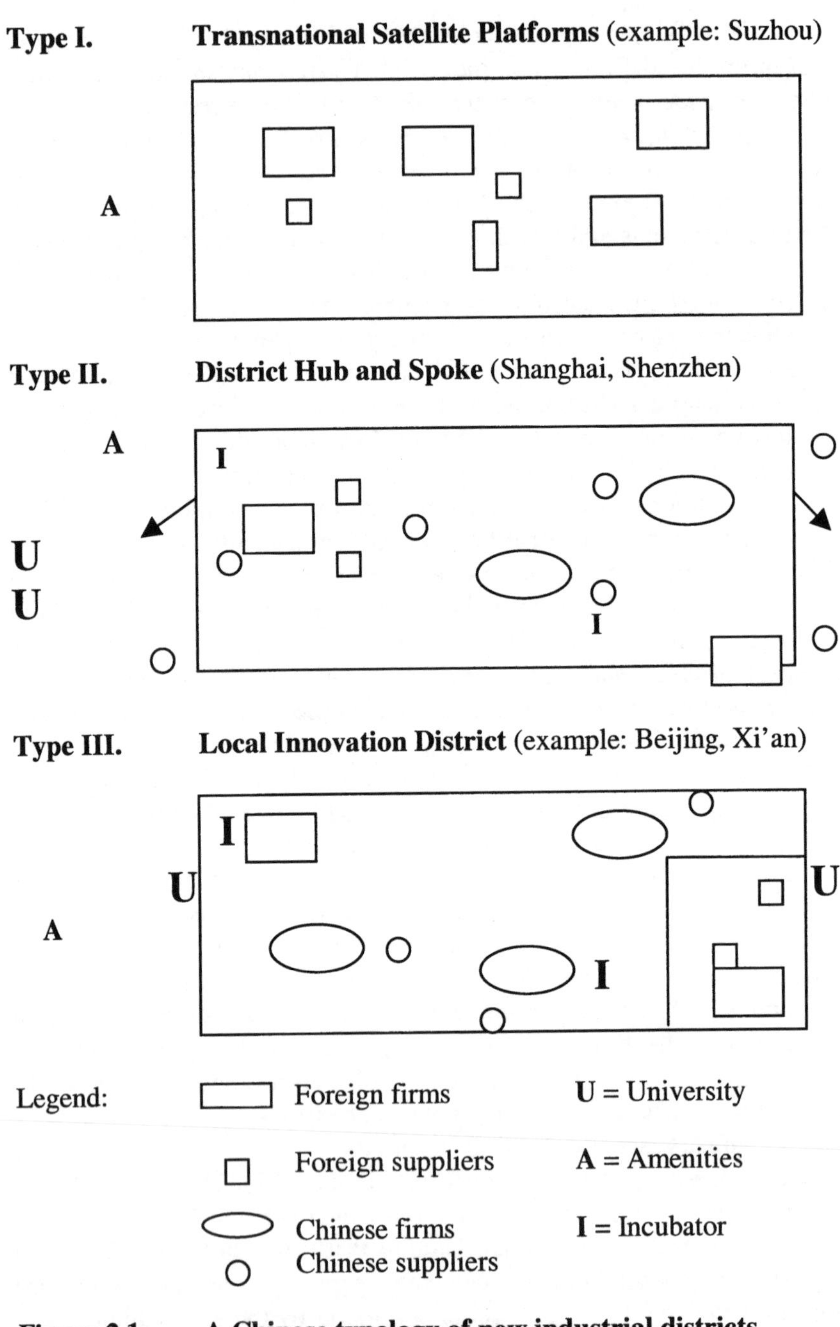

Figure 2.1 A Chinese typology of new industrial districts

The second form is a district-based modified hub and spoke. The surrounding region is the setting for spoke supplier firms, with mixed local and transnational corporate activity. The District Hub and Spoke model maintains some university connections through transportation lines for imported high skill labor. It is on the outskirts or within the town. The third model is a Local Innovation District, distinguished by the proximate presence of large R&D intensive universities and higher education institutions. It is also on the edge of the city. Significant features include a large incubator, and parks-within-park dominated by a large research-intensive company and its associated firms.

Science parks occupy space specifically set aside and designed to attract companies incorporating higher than average amounts of research and development in their products. The goal for development district planners is to promote economic development by securing companies with high profit margins and employees with high pay who will pump more money into the local economy. Numerous case studies have assessed reasons for the relative success or failure of these schemes (OECD, 1987; Castells and Hall, 1994; Camagni, 1995; JV/SV, 1995; Lyons and Luker, 1998). While it is widely acknowledged that access to cutting edge information concerning product and process development is the key to competitive success of both product and place where produced, the magic recipe remains elusive even as the model is continually reproduced on an increasingly global basis. As corporations extend their markets and production sites throughout the world, countries seek to accelerate their own development by constructing spaces attractive to desirable 'high technology' operations. These often involve companies whose highest tech components remain in their headquarters country, but whose level of R&D infusion exceeds that prevalent in the host country. This development differential is the eagerly sought object of local technology transfer. But how well will what works in the West succeed in Shanghai, or Xi'an?

It is hypothesized that the main goal of science park development in the eyes of the host geographical location is to increase the number of better paying, higher-level jobs, in addition to conveying knowledge more relevant to future development of production processes. It is further hypothesized that the strategies of areas seeking to secure these benefits are also the same, since all are dealing with capitalist companies in similar sectors. The political, cultural, and economic context, however, differs. This research therefore explores what differences are necessary due to contextual exigencies and which constitute remediable inefficiencies.

Firms reflect their utilization of territory specific advantages, which is why the geography of the location decision remains critically important, but this relationship is not deterministic. Several factors are territorially bound, relative to others: the type (skill level, training, attitude) and amount of labor available, incentives provided, technology spillovers, are all spatially constrained to a particular area. These R&D spillovers involve non-remunerated tacit (face-to-face), uncodified learning that conveys great product value in a tightly geographically bounded area (Baptista, 2001). The 'atmosphere' or historically reinforced culture of a place creates 'relational assets' (Storper, 1997) that can be used to a place-bound firm's advantage – or disadvantage (Storper, 1992). Circumstances create possibilities, but individuals act on them – or not. Human networks can be constructed to overcome or maximize territorial boundaries, generating much literature on the self-help nature of this important tool.

Geographic studies of science parks generally divide between critics who think they're a wonderful economic development idea (Bower, 1992; Braun and McHone, 1992; Burritt, 1999) and those who deride the notion as wishful thinking at best and misappropriation of public funds for private gain at worst (Massey et al., 1992; Massey and Wield, 1992). Designation and construction of technology parks undeniably takes the form of a 'property initiative' supporting (usually by government incentives) a change in use of property for a particular purpose. The property can be wholly privately owned, acquired from private hands by a not-for-profit, public-private entity involving some public funds, or be wholly a government endeavor so both problems and profits fall on public shoulders. They are usually specialized in function, encompassing administrative offices and research and development, with manufacturing facilities located on less expensive land due to size and infrastructure requirements. Such an arrangement is much more of a practical than socially constructed hierarchic exclusionary nature, however (Massey and Wield, 1992).

Success of development parks can be measured by their effect of increasing local employment through attraction of new firms, particularly those engaged in relatively high research inputs and involving links between local academics (and their institutions) and local businesses (Massey and Wield, 1992). Whether or not science parks are over-sold by their promoters as innovation centers of wholly new discoveries rather than business-nurturing sites are also beside the point of assessing their true spatial contribution. The larger disciplinary theme concerns the role of proximity and agglomeration in high tech corporate networks, and the ability to draw on local pools of diverse labor markets. Another application concerns the cultivation of regional concentrations of high tech endeavors on the intra-regional scale.

Agglomeration Networks

Examinations of spatial clustering reflect core issues in economic geography: the effects of space, place, economics, and proximity (Malmberg and Maskell, 2001). An understanding of the mechanics behind this process holds great policy implications for both political and business entities. Unfortunately, much of the information accumulated is descriptive, based on the responses of people in pre-selected places rather than systematic and able to be repeated under identical conditions. What can be said about agglomerations applies to what exists at a particular place and time, but corresponds to some general conditions that are globally verifiable.

The spatial concentration of firms and individuals related in the production process is known as an agglomeration. This phenomenon is seen as economically desirable, since reduction of spatial separation presumably leads to an increase in the transmission of information, which is then presumed to lead to an increase in innovative activities and a decrease in the learning curve, thus creating the most efficient and desirable spatial configuration. Knowledge creation needs targeted and specific investments to produce R&D activity leading to the transformation of discovery output into a marketable product. This process includes low-technology activities in firms at the stage of adapting and refining known processes and products to conform to locally appropriate conditions. Low technology activities often take the form of assembly activities for high technology products – with the R&D invested elsewhere.

Geographers have proposed a variety of theories directed at the relationship between the location of an economic entity and its performance. They pay particular attention to dynamics of the location decision and subsequent agglomeration effects of clustered industries. Agglomeration theories generally divide into five camps. The first relies on growth center/pole dynamics feeding off economies of scope to benefit from transaction economies resulting from spatial affiliations of localized backward (supply) and forward (market) linkages (Scott, 1982; Scott and Storper, 1987). A second approach prioritizes product cycle dynamics utilizing the availability of specialized labor market elements and mixes needed at a particular time in the product development cycle (Vernon, 1966). The internal organization of an industry, corporate policy toward personnel and technology procurement, or the particular social division of labor provide a competitive advantage that draws a collection of companies to a certain site (Scott, 1988). A third school looks at product type and the organizational structure of firms (Glasmeier, 1987). The fourth approach looks at the combination of history and policy by the local milieu (Malmberg, 1996). The last approach proposes the best fit needs an 'eclectic mix' utilizing elements of several of the above to supplement locally missing pieces (Hall, 1990; Sternberg, 1996).

Two concepts frequently used in association with agglomeration mechanisms need to be differentiated, as they are occasionally conflated erroneously. 'Urbanization economies' derive from the advantages to a company from locating in or close to a city (Hoover, 1937). The urban setting supplies extensive and affordable services that are scarce elsewhere. 'Localization economies' reflect the financial (and in high technology firms, the information related) advantages of firms that are located closely together. These factors often reinforce each other, since most clusters occur near a city.

Since Marshall's (1890) pioneering work over a century ago at the dawn of the Industrial Revolution, the notion has become generally accepted and widely illustrated that companies located in close proximity (a.k.a., an agglomeration) experience more rapid growth than those in separate areas. More recent observations of the transmission process of innovations led to insights of a sociological nature, that more than quantitatively rendered cost-and benefit 'factors of production' are involved (Asheim, 2000). Traditional Heckscher-Ohlin factor endowment models explained trade between nations as a cost-benefit calculation to obtain immobile but needed commodities. Industrial districts, like other geographically designated areas sharing a name, are not all alike. Inclusion of 'hi-tech' or 'science' in the name reflects more the hopes of the designator than a proportionate share of the reality of activity taking place within the park. Industrial districts come in various geographic guises, from the Marshallian classic, to the Italianate association of village or family ties, to the neo-Marshallian global node of dispersed sites (Amin and Thrift, 1992). The type of Chinese industrial districts examined in this work do not fall into the Marshallian type, since their success lay in spatially concentrating the full scope of the necessary division of labor and product differentiation needed in a concentrated industry production chain. Nor are they purely an 'innovative milieu' (Camagni, 1995) devoted to a localized learning process enhanced by tacit (face-to-face) transmission of uncodified (not written down) knowledge between trusting business associates.

Porter defined industrial clusters as 'industries related by links of various kinds' (Porter, 1990, p.131). His examination of industrial clusters used a diamond factor metaphor, launching a drive by regional economic development groups to assemble some production chain-specific coterie of companies as a base for their territorial responsibility. Pressures fuelling the drive to success via innovation are internally generated pressure, not entirely amenable to externally constructed features, however, and often products of embedded, pre-existing cultural capacity of local actors as demonstrated notably by Saxenian's contrast of Silicon Valley and Boston's Route 128 for nurturing technology firms (Saxenian, 1992). The space-time trick consists in projecting internal structures of a successful firm across vast distances of space and culture to transfer, project and re-establish that success on a global scale. This in turn requires choices – and variations – in what features of firms in one area are retained and transferred to a new location, or allowed to fit the new context of operations.

Globalization Effects

The experiences of firms and economic development paths in East Asia particularly illustrate the inappropriateness of theorists who prematurely acclaim the demise of the nation state and assorted political boundaries in the face of economic globalization. The shrinking amount of time taken to cover distances is a significant contribution of technology, but place creates uneven spaces due in part to the differences in the availability of technology (internet linkages, port facilities, etc.) in different places. Institutions and social as well as political structures still crucially impact a nation's economic status, as well as places at the subnational scale (Lundvall and Maskell, 2000). Scandinavian and European theorists particularly like to discuss 'national systems of innovation' (Lundvall, 1985). The relationship of such systems co-existing with increasingly globalized transaction systems lies in the dynamism of unevenness upon which capitalism thrives. Indeed, localization economies and globalization spatial dispersion can be considered as two sides of the same coin. In this light, MNEs function as critical connections between profit seekers across different spatial scales, feeding growth poles within and between different countries (Dunning, 2000). The localization aspect tightly ties labor pools and plants in close spatial proximity within micro-scale regions, but attracts these plants at a global scale of advantage seeking (Scott, 1990).

Role of Networks

Many areas contain specific factors that allow the construction of competitive advantages. This can occur by either legal or political structures, locational positioning, or 'social capital' of human systems which encourages both knowledge creation and transmission. A study of European cities known for clusters of a particular industry, varying from 'cultural' to 'medical', found certain features predictive as had many other preceding studies: cluster dynamics reflect the economic and spatial conditions of each particular city (van den Berg, Braun, van Winden, 2001). Networks that underlie and sustain these clusters also reflect local and individual dynamics present. Spatially delineated clusters are well worth cultivating for the value they add via jobs and economic well being of each urban region wherein they are located.

China's organization of within and between firms' network functioning to successfully enhance competitiveness through constant improvement in a trust bound environment, is modeled by Japanese and Taiwanese advanced manufacturers of the most sophisticated products. Japanese firms actually prefer working as part of a global network, and locate accordingly to minimize uncertainty while maximizing learning opportunities (Malecki, 2000). Note the lower priority of cost factors. This comes with the recognition that technology transfer mechanisms vary according to the structures of various national systems of innovation. China's system of under-investing in fairly mobile workers resembles that of North America rather than Europe or Japan's.

Foreign Factors

Foreign firms considering places to locate in China face additional complicating considerations. They must take into account not only the usual economic factors (e.g., cost of facilities, labor, infrastructure), but also the exigencies of China's stage of economic transition: the demands and conditions of the political, economic, cultural, historical and technological infrastructure. Foreign interests often enter a new setting in gradual stages. At its simplest, a representative of the company arrives with a suitcase and sets up an office in a hotel room, cushioned from unfamiliar outside conditions by English-speaking staff that may eventually become the first local corporate employees. The next step often involves setting up a joint venture, which under Chinese law can either be an equity joint venture or a contract joint venture. Powers of each partner are reflected in the former by the proportion of capital contributed; in the latter arrangement, powers are designated in the initial contract, usually allocating more to the Chinese partner to reflect the non-pecuniary contacts and local knowledge they contribute in lieu of financial capital contributed by the foreign partner. The final step is formation of a wholly foreign owned enterprise (referred to as a WFOE, or 'woofy'). Wholly foreign enterprises concentrate along the eastern coast and the capital region, reflecting the designated zones and new industrial districts, or tech parks islands created by the government to attract them. Joint ventures tend to be more evenly distributed throughout China, reflecting the greater mobility provided by the local partners and their inclination to produce more for the domestic market rather than for export, as required by the port-hugging more heavily foreign firms (He and OhUallachain, 2001).

The benefit for both foreign companies and China lies in the function of zones as 'learning regions'. A 'learning region' is characterized by a high level of mutual learning among actors, a vast amount of knowledge exchanged leading to innovations. What is actually learned, or transferred, can vary from a one-way transmission or training from the relatively advanced firm, a mutual exchange, a commercialization partnership along a production chain, or a financial arrangement either as a straight investment or an exchange for an intangible unique factor such as personal connections (Reddy, 2000). Where partnerships do occur, the key facilitator is usually an organizational impetus (Malmberg, et al., 1996; Maskell, et al., 1998). Learning – and the accompanying competitive advantage of innovation – are social creations that are nourished through networks at various scales, from the local microcosm of an industrial district through the global reach of corporate and/or industry related networks (Hotz-Hart, 2000). Entry into a new market or production zone occurs gradually, as theorized in various steps. Exports can create a demand, followed by a small representative office to explore foreign possibilities, then a joint partnership, and finally a wholly owned facility. Various types of relationships are constructed between headquarters and their extensions, as formulated at the national (Markusen 1996) and international (Park and Markusen, 1995) scale, from branch plants to orbiting satellites and ultimately autonomous spin-offs.

High Tech parks inhabited largely by TNCs draw upon both local advantages in that place and advantages of advanced production practices and/or product characteristics held within the firm and its home network (Oinas, 2000). The degree to which these characteristics travel to other firms, or are imparted to foreign employees, will be explored in the case studies. One of the challenges for China identified previously is the paucity of evident technology transfer (Young and Lan, 1997). As this study demonstrates, however, various types of technology transfer and innovation commodification do occur, but in particular companies and places rather than as a widespread, structural phenomenon. The most widespread, quiet kind of technology transfer involves the nature and content of worker training (as part of the corporate family), while TNCs learn their new responsibility for providing worker housing and cafeteria food (Dobson and Yue, 1997). While capital tends to flow from parent company to overseas subsidiary, staying within the corporate network, China's push to nationalize TNC personnel appears quite widespread. This research explores the kinds of organizations established, in different industrial parks and with varying degrees of success, to facilitate learning experiences.

References

Amin, A. (2000), 'Industrial Districts', in Sheppard, E. and Barnes, T. J. (eds.), *Companion to Economic Geography*, pp. 149-68. Blackwell, Oxford, UK.

Amin, A. and Thrift, N. (1992), 'Neo-Marshallian Nodes in Global Networks', *International Journal of Urban and Regional Research*, vol. 16, pp. 571-87.

Asheim, B. (2000), 'Industrial Districts: The Contributions of Marshall and Beyond', in Clark, G., Gertler, M. and Feldman, M. (eds.), *Handbook of Economic Geography*, pp. 413-31. Oxford University Press, Oxford, UK.

Baptista, R. (2001), 'Geographical Clusters and Innovation Diffusion', *Technological Forecasting and Social Change*, vol. 66, pp. 31-46.

Bower, D. (1992), *Company and Campus Partnership*, Routledge, London.

Braun, B. and McHone, W. (1992), 'Science Parks as Economic Development Policy: a Case Study Approach', *Economic Development Quarterly*, vol. 6, pp. 135-47.

Burritt, (1999), A pioneer turns 40: Telecommunications competition sparks a middle-age rejuvenation, *Atlanta Journal-Constitution*, Feb. 21:D1, pp. 4-5

Camagni, R. (1995), 'The concept of innovative milieu and its relevance for public Policies in European lagging regions', *Papers in Regional Science: The Journal of the Regional Science Association*, vol. 74, pp. 317-340.

Castells, M. (1996), *'The Rise of the Network Society'*, Blackwell, Oxford.

Castells, M. and Hall, P. (1994), *Technopoles of the World: The Making of 21st Century Industrial Complexes*, Routledge, London.

Coase, R.H. (1937), 'On the Nature of the Firm', *Economica*, vol. 4, pp. 386-405.

Dicken, P. (1998), *Global Shift, 3rd edition,* Paul Chapman Publishing, London.

Dicken, P. (2000), 'Places and Flows: Situating International Investment', in Clark, G., Gertler, M. and Feldman, M.(eds.), *Handbook of Economic Geography*, pp. 275-291. Oxford, University Press, Oxford, UK.

Dicken, P. and Malmberg, M. (2001), 'Firms in Territories: A Relational Perspective', *Economic Geography*, vol. 77, pp. 345-63.

Dobson, W. and Yue, C.S. (eds.) (1997) *Multinationals and East Asian Integration*, International Development Research Center, Canada.

Gerreffi, G., Korzeniewicz, M. and Korzeniewicz, R. (eds.) (1994) Introduction: Global commodity chains, *Commodity Chains and Global Capitalism*, Greenwood Press, Westport, CN.

Glaeser, E. (1998), 'Are Cities Dying?' *Journal of Economic Perspectives'*, vol. 12, pp. 139-160.

Glasmeier, A. (1987), 'Factors Governing the Development of High Tech Industry Agglomerations: A Tale of Three Cities', *Regional Studies*, vol. 22, pp. 287-301.

Granovetter, M. (1985), 'Economic Action and Social Structure: The Problem of Embeddedness', *American Journal of Sociology*, vol. 91, pp. 481-510.

Griliches, Z. (1979), 'Issues in Assessing the Contribution of R&D to Productivity Growth', *Bell Journal of Economics*, vol. 10, pp. 92-116.

Hall, P. (1990), 'The Generation of Innovative Milieu: An Essay in Theoretical Synthesis', *Working Paper 505, Institute of Urban and Regional Development*, University of California Press, Berkeley.

He, C. and OhUallachain, B. (2001), 'Entry Mode and Location: Foreign Manufacturers In Transitional China', Paper presented at Southeastern Division of the Association of American Geographers' Convention.

Henderson, J., Dicken, P., Hess, M., Coe, N. and Yeung, H. (2002), 'Global Production Networks and the Analysis of Economic Development', Global Production Network Working Paper 1.

Hoover, E. (1948), *The Location of Economic Activity*, McGraw-Hill Book Company, New York.

Hotz-Hart, B. (2000), 'Innovation Networks, Regions, and Globalization', in G. Clark, M. Gertler and M. Feldman (eds.), *Handbook of Economic Geography*, pp. 432-50.Oxford University Press, Oxford, UK.

Lundvall, B.A. (1988), 'Innovation as an Interactive Process: From User-Producer Interaction to the National System of Innovation', in Dosi, G., *Technical Change and Innovative Theory*, Pinter Publishers, London.

Lundvall, B.A. and Maskell, P. (2000), 'Nation States and Economic Development: From National Systems of Production to National Systems of Knowledge Creation and Learning', in Clark, G., Gertler, M. and Feldman, M. (eds.), *Handbook of Economic Geography*, pp. 353-72. Oxford University Press, Oxford, UK.

Lyons, D. and Luker, B. Jr. 1998. Explaining the Contemporary Spatial Structure of High Tech Employment in Texas. *Urban Geography 19*, pp. 431-458.

Malecki, E. (2000), 'Network Models for Technology-Based Growth', in Z. Acs (ed.), *Regional Innovation, Knowledge and Global Change*, pp.187-204. Pinter, London.

Malmberg, A. (1996), 'Industrial Geography: Agglomeration and Local Milieu', *Progress in Human Geography*, vol. 20, pp. 392-403.

Malmberg, A. and Maskell, P. (2001), 'The Elusive Concept of Localization Economies: Towards a Knowledge-based Theory of Spatial Clustering', Paper presented at Association of American Geographers Annual Conference, New York.

Malmberg, A., Solvell, O. and Zander, I. (1996), 'Spatial Clustering, Local Accumulation of Knowledge and Firm Competitiveness', *Geografiska Annaler*, vol. 78B, pp. 85-97.

Markusen, A. (1996), 'Sticky Places in Slippery Space: A Typology of Industrial Districts', *Economic Geography*, vol. 72, pp. 293-313.

Marshall, A. (1890), *Principles of Economics, 8th edition,1920*, Macmillan, London.

Martin, R. (1999), 'The New Economic Geography of Money', in R. Martin (ed.), *Money and the Space Economy*, John Wiley, New York, pp. 3-27.

Maskell, P., Eskelinen, H., Hannibalsson, I., Malmberg, A. and Vatne, E. (1998), *Competitiveness, Localized Learning and Regional Development – Specialization and Prosperity in Small Open Economies*, Routledge, London.

Massey, D. (1994), *Space, Place and Gender*, University of Minnesota Press, Minneapolis, MN.

Massey, D., Quintas, P. and Wield, D. (1992), *High Tech Fantasies: Science Parks in Society, Science and Space*, Routledge, London.

Massey, D. and Wield, D. (1992), 'Science Parks: A Concept in Science, Society, and 'Space' (A Realist Tale)', *Environment and Planning D: Society and Space* vol. 10, pp. 411-22.

Oinas, P. (2000), 'Distance and Learning: Does Proximity Matter?' In Boekema, et al., (eds.) *Knowledge, Innovation and Economic Growth: The Theory and Practice of Learning Regions'*, pp. 57-69, Edward Elgar, Cheltenham, UK.

Organization for Economic Co-operation and Development (1987), *Science Parks and Technology Complexes in Relation to Regional Development,* OECD, Paris.

Park, S. O. (1996), 'Networks and Embeddedness in the Dynamic Types of New Industrial Districts, *Progress in Human Geography*, 20 (4) pp. 476-493.

Park, S.O. and Markusen, A. (1995), 'Generalizing New Industrial Districts: A Theoretical Agenda and an Application From a Non-Western Economy', *Environment and Planning A*, vol. 27, pp. 81-104.

Piore, M. and Sabel, C. (1984), *The Second Industrial Divide*, Basic Books, New York.

Porter, M. (1990), *The Competitive Advantage of Nations*, Macmillan, London.

Porter, M. (2000), Location, Competition, and Economic Development: Local Clusters in a Global Economy, *Economic Development Quarterly*, vol. 14, pp.15-34.

Reddy, P. (2000), *Globalization of Corporate R&D: Implications for Innovation Systems in Host Countries*, Routledge, London.

Romer, P. (1990), 'Endogenous Technological Change', *Journal of Political Economy,* vol. 98, pp. 72-102.

Saxenian, A. (1992), 'Contrasting Patterns of Business Organizations in Silicon Valley', *Environment and Planning D: Society and Space*, vol.10, pp. 377-91.

Scott, A. (1982), 'Production System Dynamics and Metropolitan Development', *Economic Geography*, pp. 185-200.

Scott, A. (1990), 'The Technopoles of Southern California', *Environment and Planning A*, vol. 22, pp. 1575-1605.

Scott, A. and Storper, M. (1987), 'High Technology Industry and Regional Development: A Theoretical Critique and Reconstruction', *International Social Science Journal*, vol. 39, pp. 215-32.

Sheppard, E. (2001), ' The Spaces and Times of Globalization: Place, Scale, Networks and Positionality', Paper prepared for 'Geographies of Global Economic Change' Conference, Clark University, Worcester, MA.

Sternberg, R. (1996), 'Regional Growth Theories and High-Tech Regions', *International Journal of Urban and Regional Research*, vol. 20, pp. 518-538.

Storper, M. (1992), 'The Limits to Globalization: Technology Districts and International Trade', *Economic Geography*, vol. 68, pp. 60-93.

Storper, M. (1997), *The Regional World: Territorial Development in a Global Economy*, The Guilford Press, New York.

Taylor, M. and Asheim, B. (2001), ' The Concept of the Firm in Economic Geography', *Economic Geography*, vol. 77, pp. 315-28.

Thrift, N. and Olds, K. (1996), 'Refiguring the Economic in Economic Geography', *Progress in Human Geography*, vol. 20, pp. 311-37.

Van den Berg, L., Braun, E., and Winden, W. (2002), *Growth Clusters in European Metropolitan Cities*, Ashgate, Aldershot, UK.

Vernon, R. (1966), 'International Investment and International Trade in the Product Cycle', *Quarterly Journal of Economics*, vol. 80, pp. 190-207.

Westhead, P. and Batstone, S. (1998), 'Independent Technology-based Firms: The Perceived Benefits of a Science Park Location', *Urban Studies*, vol. 35, pp. 2197-219.

Williamson, O.E. (1983), 'Organizational Innovation: The Transaction-Cost Approach', in J. Ronen, (ed.), *Entrepreneurship*, Heath, Lexington, MA, pp. 101-34.

Yeung, H. (1998), 'The Social-Spatial Constitution of Business Organizations: A Geographical Perspective, *Organization*, vol. 5, pp. 101-28.

Chapter 3

Chinese Policy Background

The Chinese Model

Spatially restricted zones for foreign economic activity in China were established over a century and a half ago, notably in the case of Hong Kong island which was transformed from a marginal agricultural outpost on the Pearl River delta downriver from Guangdong (then known as 'Canton' in the West) to a bustling global trade entrepot. The city is now a re-incorporated economic engine for South China. Its first modern counterparts since Deng Xiaoping's 'opening and reform' in the late 1970s are the *Special Economic Zones* (SEZ), established initially in Shenzhen, Zhuhai, Shantou and Xiamen, all in south China near Hong Kong. Hainan island, another peripheral southern area (therefore relatively safely removed from the locational main stream of China) became a SEZ in 1988. Other spatial zones designed to attract foreign investment include *Open Coastal Cities*, confined to 14 south and east coast municipalities when they were created in 1984. All provincial capitals were granted Open City status in 1992. Pudong New Area, directly east of central Shanghai, became a 'super SEZ' when it was designated for fast track development in 1990 (Yeh, 1996; Yusuf, 1997).

More specialized areas with the particular mission of fostering businesses in the technology-intensive sector of industry are the 54 *Economic and Technological Development Zones* (ETDZ). Between 1984-1988, the State Council approved 14 of these designated areas within the open cities. They were largely created on greenfield sites adjacent or proximate to established cities to take advantage of urbanization economies. Additional ETDZs were established between 1992-1993. China-Singapore Suzhou Industrial Park (CSSIP) and Shanghai Jinqiao Export and Processing Zone fall into this category, despite the name disparity. The chief difference between this type of zone and the High Tech industry zones are the emphasis of the latter, yet high technology companies locate in both (www.moftec.gov.cn).

China's 15 (as of mid-2002) *Free Trade Zones* primarily feature tax-free bonded warehousing, expedited and less expensive import and export of parts and products. They are all located in harbor cities along the east coast, and include functions expediting duty free trade, export processing, logistic warehousing and bonded commodity displays. Shanghai's Waigaoqiao Duty Free Zone is the country's oldest (June 1990) and largest. Since April 2000, the State Council approved 15 *Export and Processing Zones*, often located within ETDZs. Examples of this overlapping of categories include Shanghai's Songjiang Export and Processing Zone near Minhang Economic and Technological Zone, and Dalian's Export Processing Zone opposite Dalian's ETZ. Tianjin, the major manufacturing port city closest to Beijing, contains three separate development zones.

The 53 national-level *Science and Technology Industrial Parks* (STIPs) represent a further refinement of this notion. They seek to take advantage of the short-distance 'spillover effect' of technology by locating close to a university or research site. They form the principal geographic locus of examination for this research. Various levels of government and private entities sponsor a variety of other zones established to attract businesses to their locale, but without the imprimatur and approval of the central government (Economist 2002). Beijing's Zhongguancun pioneered the model in 1988, followed by an additional 26 STIPs in 1991 and 25 more 'Hi Tech Industrial Development Zones' in 1992. The last addition was Shaanxi's Yangling Agriculture Development Zone in 1997. The local names vary by designation, but the 53 are considered a group with similar characteristics in official state documents. Ten of these zones were further granted the distinction of being open to Asia-Pacific Economic Community (APEC) members in 1997. The Ministry of Science and Technology and the Ministry of Education jointly approve an additional 22 *University Science and Technology Parks*. Nearly every university with a strong tradition in technology-based research is encouraged to construct a park close to or within the campus as a location for indigenous technology transfer to a business application.

China's creation and promotion of science and technology industrial parks (*Kejing Gongye Yuan*) form an important part of the country's strategy for taking its place as a major and respected player in the global economy.

These spatially designated areas are the incubators for nurturing companies in favored high technology sectors that can accelerate China's development. Speculation about possible outcomes falls into various metaphors. Will technology transfer from foreign firms, advanced domestic research institutions, and student entrepreneurs returned from abroad enable China to 'leapfrog' into a more developed status, instead of falling into formation behind the Japanese-led 'flying geese' migration as have other Asian 'tiger cubs' (Feinstein and Howe, 1997)? Or is the only practical way to continue 'crossing the river by feeling for stones' and inventing incremental 'Chinese characteristics' for developing this large and resource-rich country? A variety of terms in Chinese are used to describe the kind of firm finding popular approval under the 'High Tech' (*Gao Ji Shu*) rubric.

Four factors distinguish the current Chinese development district model from previous patterns in developing countries. The first is China's size, both its land mass and more importantly its large population. A huge pool of hard-working, low pay scale, semi-educated workers supply an enormous workforce by migrating into job opportunity regions from the interior. This situation distinguishes China from the less populous Asian Tigers. The labor base in China includes a highly educated segment capable of innovating but hampered by a shortage of both investment capital and trained management. China's access to models of development to study and choose from constitutes the second distinct factor. Like Japan at the initial stage of Western influence a century and a half ago, China actively sends agents abroad to acquire knowledge about such models in order to construct its own path for modernization. At the same time, a third distinction from other Asian models emerges with China's lack of a strong, wealthy mentor country, due to confining political allegiances in the communist bloc. The fourth factor is the unfolding of a transitioning state from economic centralization to increased localization of control, within an overall framework of socio-political-military controlled stability. New economic initiatives and industry support come from central political authorities. As in other developing countries, government entities play major roles in the development of targeted localities.

High Technology Industries Definition

Over 10,000 companies currently fall into China's definition of high technology, which rests on four pillars (Wang, 2002):

- Electronics and information
- Optics-machinery-electronics integration
- New materials
- Combined bioengineering and pharmaceuticals

This list of favored technologies for development also includes:

- Advanced manufacturing technology
- Aeronautical and aerospace
- Marine engineering
- Nuclear utilization
- New sources of energy and high efficiency energy conservation
- Environmental protection
- Agricultural technology
- New technologies and processes applied to updating traditional industries (a very broad category!) (www.chinagate.com.cn 8/9/02)

China's determination to develop its own high technology base was rekindled in 1990 by two events: the fall of the former Soviet Union (its major technology supplier and on-again-off-again mentor since 1949) and the technological prowess displayed by the U.S. in the military subduing of Iraq. The commitment to advance developments in computers and other high technology fields is prominent in every five year plan since 1990, with some fruits of that policy already evident. Computer production soared from negligible in 1990 to 1.9 per cent of world production in 1995 and 6.9 per cent in 2000 (McMillion, 2002. This reflects in part China's integration into the global economy as a low cost assembly site, but also includes figures for production of China's own 'Legend' brand computer. It also indicates the increased opportunities for Chinese industry to learn from foreign computer and parts manufacturers. The drop in technology exports from Hong Kong to the U.S, and the corresponding rise in exports to the U.S. of these products from China in the 1990s, indicates the geographic shift in the location of production from the former colony to the motherland.

Initial Post-Mao Technology Policy Shifts

Premier Deng Xiaoping's statement during a National Science Conference in March 1978 that 'science and technology are the primary productive forces' signaled major policy realignments with broad spatial ramifications (Simon and Goldman, 1989). The proclamation of a 'Reform and Openness' policy by the 11th Chinese Communist Party Congress, in December of the same year, marked an historic shift in emphasis from class struggle to economic development as China's central ordering principle. The establishment of special zones for accelerated industrial modernization dates from this time forward, beginning with the eastern coastal areas. Historically, these sections of the country possessed the most potential due to longstanding ties with advanced world regions and the presence of the country's premier research centers.

Policies to promote increasing the technology level of industries in China fall into the Asian pattern of strong government participation, particularly at the national level. Provincial and municipal political units also participate heavily where they choose (and are powerful enough) to become involved, to a much greater extent than is prevalent in the West. One example frequently exercised is the central government's proclivity to pick geographically specific sites for targeted economic assistance. The first Special Economic Zones (SEZs) were established in 1979 in four locations chosen for their proximity to sources of investment capital close to but outside the PRC: Shenzhen, Shangtou and Zhuhai near Hong Kong, and Xiamen near Taiwan. The national government also funds and directs faculty and university graduates to special laboratories (the top tier designated as 'Key Laboratories') and research centers.

Within one year from proclamation of the new science and technology policy, the number of research-oriented institutes, funding, and personnel had increased greatly. In 1980, a professor affiliated with Beijing's Chinese Academy of Sciences (CAS) established the first successful individual commercial enterprise from a laboratory technology transfer (Gu, 1996). Within five years, this entrepreneurial venture grew to around 100 similar businesses, particularly along what became known as 'Electronics Street', in the *Zhongguancun* district close to Peking and Qinghua University, both scientific research and engineering centers (Wang and Wang, 1998; Wang, 1999). Emphasis on developing consumer rather than military focused technology represented both an important policy reorientation and practical recognition of China's need to attract foreign investment at a level that it could absorb, given its shortage of capital and supportive infrastructure. The success of 'Electronic Street' spawned an endless, unwarranted number of claims throughout China that every major town could concoct its own 'Silicon Valley' (Zou, 2000).

The Toffler's book on 'Future Shock' caught the imagination of China's State Council by 1983, coinciding with a forum on 'The challenge of world new-tech revolution and our countermeasures' (Wang 2000). The military preparedness metaphor still captured China's imagination and attention. 'Open Cities' along the eastern coast and 'Economic and Technology Development Zones' (ETDZs) were created in 1984 to extend the spatial extent of preferred development areas and focus foreign investment in particular sites. In July 1985 the Shenzhen municipal government and CAS founded the 'Shenzhen Science and Technology Industrial Park' (Yang, 2000). This marked the first spatially delineated attempt to lure foreign investment for Chinese high technology development in a specially developed area. Four years later the municipal government established Shenzhen Hi Tech Industrial Park (SHIP), subsuming the original industrial park into one of 27 State science and technology industrial parks (www.chinatorch.com 1999). SHIP constituted the first district to combine research from institutions affiliated with the Chinese Academy of Science and foreign company technology transfers.

A 'Decision on Reforms of the Science and Technology System' promulgated in 1985 led to better financial support, more autonomy, and a stronger focus on promoting economic links for research institutions. By increasing marketization of research output, the government sought to make scientists more responsive to commercial needs for products that could realize profits in an increasingly capitalist system. At the same time, to promote innovative applications of appropriate technology in another segment of the labor pool, the government launched the 'Spark Program' targeted toward enhancing the productivity of rural industries (Qu, 1999). This program proved particularly helpful for providing capital and advice to township and village enterprises (TVEs), discussed more fully in the Dongguan case study. Encouragement of TVEs tied into the development of a new spatialization of the production chain, forging tighter links between towns and cities as heavy industries were relocated from dense urban populations to outlying suburbs. The smaller scale TVEs frequently served as parts suppliers for decentralized large state owned enterprises (SOEs) such as Baoshan Steel, China's largest foundry on the outskirts of Shanghai.

The '863' Plan

The first national level government program specifically providing capital for technologically promising businesses was the '863' plan, named for the date (March, 1986) of a speech launching the idea which was subsequently approved by the State Council. The impetus resembles Vannevar Bush's post-World War II efforts culminating in the establishment of the U.S. National Science Foundation in 1950 to support applied research. Fueled by concern over a perceived collusion between the U.S. and Taiwan against Mainland China, the focus fell on military projects as well as those to improve the people's livelihood. Research institutions in strategic locations such as Xi'an, in interior regions developed for a fall-back 'Third Front' (Naughton, 1988) in the event of possible military attack on China's borders, particularly benefited from this program, as did Shanghai due to its previous links to military production (Ravenhill, 1998).

Launched by the Chinese Academy of Science and Peking University under the auspices of the State Science and Technology Commission, '863' advocated government preparation of a well-endowed fund to assist high technology project research that could lead to usable products. Funds in the '863' program largely went to projects solicited from major university scientists showing some commercial 'fruits' of their ideas. Areas identified for special support included projects promoting new approaches in (listed alphabetically) automation, biotechnology, energy, information technology, lasers, new materials, and space technology. A move to secure more funds for research with commercial possibilities led the following year to a limited number of mergers of R&D entities with established firms. This did not prove particularly successful, for a variety of cultural as well as commercial and reasons (Gu, 1999). Bureaucratic rigidities in state owned enterprises later proved fatal; their inability to digest more technologically nimble operations, thus reforming from within, was symptomatic of larger organizational ills.

Currency exchange adjustments from 1986-87, spelled out in the Plaza Agreement, assisted the creation of an economic phenomenon termed the 'China Circle' (Naughton, 1997) to describe the manufacturing and trade interrelationships of Taiwan, Hong Kong (the central player), mainland China (principally the contiguous provinces of Guangdong and Fujian), and Southeast East. Since well over half of all FDI funds invested in China are generally estimated to come from Taiwan and Hong Kong, and such linkages were officially forbidden at the time and until quite recently, such a documented relationship clearly indicates the strength of economic, rather than political, imperatives since the demise of Mao – and the ability of China to attract and absorb foreign enterprise capital.

Interregional manufacturing networks in Asia date back at least to the forced relations under Japan's 'Co-Prosperity Sphere' affecting Taiwan, Korea, northern China and the home islands during World War II. Changing relative factor costs within Asia of land, labor and capital in the mid-late 1980s encouraged revisiting the cooperative possibilities of shifting geographies (Ravenhill, 1998). Linkages now run through the economic exigencies of production chains, rather than (and often in spite of) politics.

The 'Torch' Program

Many Chinese policies promoting high technology enterprises were promulgated in 1988. The Beijing Experimental Zone for the Development of New Technology constituted that city's first formal new industrial district, built largely around information technology innovations – the key sector launching Asian technology development (Boulton and Kelley, 1999). Several city blocks surrounding 'Electronic Street' were incorporated into this enterprise-promoting area. Other 'experimental zones' were created to sustain new research spin-offs, such as one bordering Fudan University in Shanghai, across a busy boulevard from the main campus. Torch recipients were initially geographically targeted for urban areas with pools of appropriately trained skilled workers and an environment open to industries and ideas from outside China, in order to facilitate technology transfer. The key hope of the Torch Program was for it to serve as a government-provided bridge between the previously isolated domains of academia and business activity.

The 'Torch' Program, launched by the Ministry of Science and Technology in August 1988, extended the central government's support to focus on small and medium size businesses. With a lower requirement than '863' for demonstrated success of an idea, but still involving initial screening by a panel of scientists, Torch encourages a broader base of applicants. Most funds are distributed for high technology projects throughout the country, combining with other venture funding to sustain small but worthy efforts. Torch investments target the second stage of company capital needs following some preliminary development launch. Since its establishment in 1988, the Torch Program has supported around 25,000 projects, including 1,358 in 2001. Principal sectors funded were projects in information technology, biological engineering, materials science, and environmental protection. The target group of small and medium enterprises represents over 80 of projects funded, according to the 2002 State Torch Plan (www.chinagate.com.cn 8/9/02).

Constructing a viable business plan is a key goal of Torch advice, particularly aimed at attracting foreign investors leery of the prevalent 'Just Do It' approach of many novice entrepreneurs. The government seeks to educate aspiring owners of high technology ventures on the value of including plans relating to the entire product development cycle, from research through development, production, marketing and distribution. The 'carrots and sticks' go beyond outright grants to include tax breaks and easier licensing regulations as inducements. Requirements for capital on hand and qualifications for business managers and scientific consultants help to ensure the viability of the enterprise proposed. Intellectual property protection remains a weak link, with leaks and rival product development continuing.

The well-founded stringency of these requirements promoted affiliation of enterprises with established R&D institutions such as the Chinese Academy of Sciences and universities. These 'parents' supplied necessary ingredients such as financial capital, physical surroundings, scientific personnel, advice on various aspects from technical to managerial, access to power networks and prestige through institutional affiliation. Such useful ties prove difficult to terminate, and often endure in partially institutionalized form, extending the time period for institutional dependency beyond physical relocation. Although exceptionally successful firms take the step of offering their stock to public investors, they often retain many ties with their parent research institution despite increased independent financial backing.

Incubator Programs

Facilities to provide maximum support for new enterprises soon after their inception are known as 'incubators' in the West and as 'innovation centers' in China (translated as 'hatcheries' in Chinese). First set up as an extension of the Torch Program, the next center was established in the new Shenzhen High Tech Park section in the new town of Shenzhen across from Hong Kong. Of over 100 innovation centers in China, the central government followed its plan of directing resources to a limited list of key colleges and universities by specifying ten top incubators including those in Beijing, Shanghai, Suzhou, and Xi'an. These cities also run special International Business Incubators (IBI). Facilities are distinguished from those in standard locations such as Dongguan, which feature a gray concrete barracks-like building with metallic furniture and serviceable computers. Other IBIs include comfortable lecture rooms, reception areas, comfortable chairs, and ample product display areas as well as receptionists. The most basic facilities are found in college dorms set aside for students who have been granted a year's leave from their studies in order to develop an idea into product form, and apply for further grants in support of its development. Their proliferation and steady success can be seen in Table 3.1.

Table 3.1 Chinese high technology incubators, 1994-2000

Category	**1994**	**1995**	**1996**	**1997**	**1998**	**1999**	**2000**
Incubators	73	73	80	100	100	110	131
Companies	1,390	1,854	2,476	2,670	4,138	5,293	7,693
Av. per incubator	19	25	31	27	41	48	59
Employees	NA	NA	NA	45,600	68,975	91,600	128,766
Av. Employees	NA	NA	NA	12.1	16.7	17.3	16.7
Total income (billionUS$)	**1.48**	**2.42**	**3.63**	**4.08**	**6.07**	**9.58**	**17.88**
Av. income (millionUS$)	1.06	1.31	1.47	1.53	1.47	1.81	2.32
#graduates	NA	174	284	177	499	618	836

Source: Harwit, 2002

Reflecting 1999 figures, the 'China Daily' government newspaper reported that a total of 300 incubators in the STIPs have graduated 1,934 new companies, representing 164,000 jobs. Their future, linked to capital constraints from conservative government financing and the lack of the less risk-averse capital from private investors prevalent in the West, remains an open question. Occupants of China's incubators are predominantly computer related start-ups, the same first success sector as in the West. If more of them can survive and profit long enough to invest funds in the next wave of entrepreneurs, like the Legend Group and Founders, the picture will be more optimistic.

Hi Tech Industrial Development Zones

Government sponsorship for 53 nation-wide 'hi tech industrial development zones' (Wall and Yin, p.172) came out of the preceding movements to meld research and commercial ventures (Table 3.2).

Table 3.2 Hi Tech parks, by technology-related sales revenue

Location	Tech-related Sales Revenue (in RMB)
* ****Beijing**	16,308,165
*Guangzhou	2,960,132
**Chengdu	2,441,072
* ****Shanghai**	1,840,269
Qingdao	1,469,228
Tianjin	1,388,304
* ****Xian (2)**	1,202,651
*Changchun	1,123,976
Shenyang	1,102,724
**Wuhan	1,057,058
*Hangzhou	858,593
*Nanjing	841,922
Wuxi	789,231
*Changsha	660,377
Foshan	577,824
*Dalian	550,697
Nanning	533,865

Haerbin	452,763
Jinan	397,407
Luoyang	347,897
Taiyuan	347,741
Zhengzhou	213,926
Jilin	194,402
*Fuzhou	166,498
* **Hefei	164,000
Baotou	162,123
Shijiazhuang	142,749
Xiangfan	141,062
Weifang	137,557
Wulumuqi	135,051
Nanchang	124,809
Hainan	82,821
****Shenzhen**	77,122
Guilin	51,508
Kunming	45,492
Baoji	41,947
Changzhou	30,881
Anshan	27,277
Baoding	25,520
Zhuzhou	24,542
*Xiamen	22,040
Guiyang	18,354
****Suzhou**	16,279
Mianyang	4,828
Zhongshan	4,343
Zibo	2,157
Xhuhai	1,600
Huizhou	1,342
Weihai	0
**Yangling	0

* Includes a 'National Torch Software Development Base'
** APEC Science and Technology Industry Park
Source: 'China Statistical Yearbook on Science and Technology' (2001)

The initial 26 'State-level new and High Tech development zones' were established in 1991. Another 25 designated areas were approved the next year. Yangling's agriculture oriented zone was added in 1997. As measured by revenue related to technology intensive categories designated by the Chinese government, they performed as shown by Table 3.2 in the year 2000.

Even as a single city, Beijing (population 11,077,300) clearly dominated this category, followed by Guangzhou (7,006,900), profiting by propinquity to Hong Kong. Interestingly, the inner China city of Chengdu (10,133,500) edged out the Yangtze delta city of Shanghai (13,216,300). In 2002, the year of this census, 40 Chinese cities had populations in excess of 1 million (China Business Review 2002). A note on these numbers: there are various agencies that keep statistics in China. What they count varies, and categories are often not comparable. These statistics are given to provide a sense of relative standings, rather than veracity in numbers.

Their success in attracting capital and producing goods is demonstrated in Table 3.3. By the year 2000, 20,796 firms were reportedly in STIPs. The 560,000 Chinese scientists working in these companies included 9,358 PhDs, 52,103 with Masters degrees, and over 5,600 returned after at least one year from some higher education institution abroad (www.chinagate.com.cn 8/9/02). Enterprises in the STIPs should have at least 30 per cent high school graduates, 10 per cent in R&D activities (Wang, et al., 1998) Almost 70 provincial level and more than 30 university level 'high- and new-tech industrial development zones' vie to incentivize their space for attracting prestigious and profitable companies in this variously interpreted classification. Software Parks are also in Hebei, Jinlu, and Yunnan. The development of high technology categorized companies can be contrasted to those in STIPs, as noted in Table 3.3. Clearly, the majority in numbers of companies and employees and value of high technology products exported are concentrated in STIPs.

China's scattered Hi-tech Industrial Development Zones encompass a wide variety of activities at each location. Subsequent chapters will explore the makeup and consequences of the parks in various parts of China selected for their leading role in development in representative parts of the country. Some, as in Beijing, are centered on

Table 3.3 Large and medium size high technology enterprises, 1991-2000

Category	1991	1995	2000
# w/tech development activities (10,000)	NA	1.3	1.0
STIP companies(10,000)			2.8
R&D personnel (10,000)	NA	26	54.3
STIP personnel (10,000)			250.9
Exports (US$ 1,000)	36,875	158,875	
STIP exports			18,581,751

Source: Adapted from CSY on S&T (2001) Hi Tech Industrial Development Zones

research institutions. Some in Shanghai are located around traditional manufacturing labor pools, or dictated in wholly new locations by government redistributive policy. Others reflect a geographical closeness to sites of overseas Chinese capital, such as Shenzhen, or military productive capacity, as in Xi'an. Local level STIP usually amount to little more than advertising for prestigious address, and principally hold manufacturing rather R&D activity. They are far less innovative than industrious, and only major universities and institutes manage to produce any significant innovative breakthroughs.

State level development areas are classified in three categories, and targeted for three types of locations, which fit well with theoretical projections of best practice location criteria for science and technology-based activities. STIPs based on R&D. They are located in large cities with major universities and research institutes, which are usually tied to political centers such as provincial capitals. This also serves to mitigate geographic unevenness, building on the strategic concerns behind Third Front development of more interior areas such as Xi'an in Shaanxi province. The second type of R&D locational consideration was to follow previously embedded traditional industrial strengths in centers with practical applications, then for the military and now for commercial ventures, though to a certain extent overlapping. Control of many economic enterprises by the People's Liberation Army (PLA) underscores the connection. A third type of location consideration is of a very different nature: the outward-looking coastal cities can draw on their previous experience of dealing with foreign enterprises and impacts to provide personnel and a place-based historical context for the special needs of again dealing with foreign enterprises and individuals. These cities range, from Haikou on southeastern Hainan Island to municipalities around southern Guangzhou, north through Ningpo and Fujian to metropolitan Shanghai on the central coast, and northeast to Tianjin and Dalian.

A series of five 'High and New-tech Industrial Development Belts' were planned in the early 1990s. Under the State Science and Technology Commission, the plan was for these initial 24 STIPs to be linked by the existing transportation and communication system. The belts, and their associated cities, were:

1. *JingJin* Belt (Beijing, Tianjin)
2. *Pearl River Delta* (Guangdong Province's Guangzhou, Shenzhen, Zhuhai, Foshan, Huizhou)
3. *Yangtze River Delta* (Suzhou, Wuxi, Changzhou, Yangzhou, Zhenjiang, Nanjing – note the early absence of Shanghai!)
4. *QiLu/Shandong Province* (Jinan, Xibo, Weifang, Qingdao, Yiantai)
5 *ShenDa /Liaoning* Province (Shenyang, Anshan, Yingdou, Haicheng, Dalian) (Wang, 2002)

Preferential policies provided by central government authorities attract the most attention from aspiring occupants, domestic and foreign. These include expedited 'one-stop' processing of forms and voluminous requirements (from the country that invented bureaucracy and the civil service), reduced rates or phased exemption from taxes depending on the age of the enterprise and its accessed degree of technology intensity. Bonded warehouses and factories are available for new High Tech firms (as featured in Shanghai's Pudong district Waigaochiao), a zone-wide import-export agency, and autonomous foreign trade business license. A category of 'National High Tech Export Bases' was created in 2000 to promote STIPs located in Beijing, Tianjin, Shanghai, Suzhou and Shenzhen for export-oriented (particularly foreign companies) businesses. These cities feature especially good port infrastructure connections and duty free zones with large warehousing capabilities for export-oriented enterprises.

To what extent firms in these new industrial districts are oriented to domestic market production, or manufacture primarily for export, reflect local technology and attract or produce capital, also varies in proportion and over time. Not every STIP produces products infused with the fruits of research, particularly IP infringement-leery multinationals that with some justification prefer to keep the key research elements in headquarter locations while modifying locally for the Chinese market. The financial and bridging functions are most important for foreign entities. For Chinese economic maturation, these islands of preferential treatment and opportunity are much more important learning districts. They remain experimental, somewhat isolated places apart from the general economy, under the auspices of the State Commission of Science and Technology (Wall and Yin, 1997). Currently, as in the past, they serve as bridging spaces between the lagging level of China's technology and capital and that of the more advanced economies, concentrating pools of China's scarce fiscal and human resources to light a spark for firing the nation's development trajectory. The goal of China's STIP promotion remains independence from economic dependence, while at the same time welcoming interdependence by inviting global companies to locate there, do business and manufacture advanced, otherwise unavailable products in China. Their success can be indicated by government figures showing that from 1991-2000, the amount of output from STIPs increased as shown in Table 3.4

Table 3.4 Increase in STIP output, 1991-2000

Category	2000 amount (billion RMB)	Increase from 1991
Technology-industry-trade income	920.9	106 times
Total industrial output value	794.2	112 times
Financial revenue	46	118 times
Export exchange earnings (US$)	18.6	103 times

Source: www.chinagate.com.cn.

Another measure of the success rates for zones in various types of configurations and locations is the amount of exports and their annual change in value, as given in Table 3.5 below.

Table 3.5 Exports by types of zones, 2000-2001

Zone	Exports (US$million)	%change from 2000
Coastal Open Cities Total	82,024.8	8.1
Shanghai	27,623.7	9
Guangzhou	11,623.6	-1.4
Tianjin	9,492.1	10
Qingdao	7,725.1	12.8
Dalian	7,416	1.6
Ningbo	6,244.2	20.8
Fuzhou	4,419.7	4.1
Nantong	1,979.2	2.2
Yantai	1,813.7	6.9
Wenzhou	1,813.5	32.3
Qinhuangdao	912.1	40.3
SEZ Total	27,932.3	5.3
Shenzhen	17,813.2	10.7
Xiamen	5,068.6	8.5
Zhuhai	3,197.9	9
Shantou	1,054.7	-48.1
Hainan	797.9	-0.6
ETDZ Total	22,612.8	14.2
New Pudong	11,510.9	20.1
Suzhou	1,678.5	8.8
Hainan	.9	10,034.9
STIPs Total	8,797	19.7

Source: Modified from Economist (2002)

The Southern Tour

Reformist Premier Deng Xiaoping embarked on a rapid trip through the economically booming southern province of Guangdong in 1992, variously termed the 'Deng Whirlwind' (Wang, 2000). He was appalled to observe that the famed independence of the southerners had reasserted itself as a byproduct of fiscal well being, exemplified by the prevalent use of the regional rather than the national dialect in the media. Guangdong's evident economic success nevertheless was held up as a national model during the 14th National Congress of the Chinese Communist Party. For the first time, he proclaimed the goal of creating a 'socialist market economy' to fuel China's drive toward rapid modernization based on technological upgrading.

That summer, the Science and Technology Commission (STC) recommended the establishment of another 20 or more STIPs – a further step for geographically diffusing the technology development funds. Approval came early the next year, by which time almost 70 provincial level STIPs and various municipal level development parks were underway without formal state-level STC oversight. Two formal 'Decisions' promoting utilization of science and technology to promote economic development were promulgated by various State bodies in 1993, 1995, and 1996. Industrial technology was clearly targeted for financial assistance and political attention.

The '973' Plan

A follow-up to the '863' Plan, the '973' Plan (advanced in March 1997) highlights nine technology projects in 'Six Key Pillar' areas including telecommunication, electronics, biotechnology, new materials, and the environment. In the teeth of economic slowdowns that hit more advanced Asian economies hard later that year, China's announced its resolve to intensify investment in high technology industries. As a further response to the regional economic crisis of 1997, China's State Council approved the designation of ten STIPs as Asia Pacific Economic Cooperation (APEC) parks. Previously established parks in Beijing, Chengdu, Hefei, Shanghai, Shenzhen, Suzhou, Wuhan, Yanling, and Yantai were singled out as particularly open to 'cooperation and exchange' between China and other APEC countries.

Picking up on a phrase that had gained currency in the West, in 1999 Deng's successor Premier Zhu Rongji called for creation of a 'national innovation system' in both tertiary as well as secondary economic enterprises. He thereby acknowledged the increasing maturation of the Chinese economy by focusing technology-enhancing efforts on the next rung of the economic development ladder, and a major arena for foreign direct investment: services.

Research Institution Restructuring

Until the push to produce marketable advanced technology in the 1990s, China's research and higher education system consisted of three branches, with several tiers. Teaching fell to the colleges and universities, with research concentrated on more theoretical topics. Applied projects were allocated by a bid or patronage system to branches of the Chinese Academy of Sciences, or special schools established under different departments of the government. From the late 1990s, this system experienced consolidation with the squeezing out of SOEs and merging of departmental institutions. At the same time, colleges and universities were encouraged to undertake applied research and spin out companies. A top tier of these institutions was singled out for government support and funding. In late 2001, 46 research institutes and 26 related manufacturing units were further merged into the China Electronic Technology Corporation by the Ministry of Information Industry and tasked with turning out globally competitive High Tech products within the following five to ten years(McMillion, 2002).

Chinese sources reported that foreign firms greatly expanded in-house R&D in the previous five years in certain industries. From 100 R&D centers in FDIs by 2000, according to MOFTEC, another 24 were added the next year according to CAS. From this author's observations, however, the nature of R&D being conducted does not involve 'cutting edge' innovations; cooperative research with Chinese universities also largely involves marketing and modification work.

Various 'institutes' and 'academies' also conduct research, under the auspices of different branches of the government (municipal, provincial, and national) and the People's Liberation Army (PLA). The latter operates as a semi-autonomous body, operating numerous commercial as well as academic entities (Pillsbury, 2002). These largely serve a policy-oriented purpose, producing studies rather than manufactured product ideas. They are more closely aligned with the political structure than are U.S. 'think tanks'.

Foreign Involvement

The pace of and need for foreign investment in China's technology development is indicated by two sets of numbers. Figure 3.1 indicates the source and level of the top foreign technology investing countries in China through the previous decade, from 1991-2000.

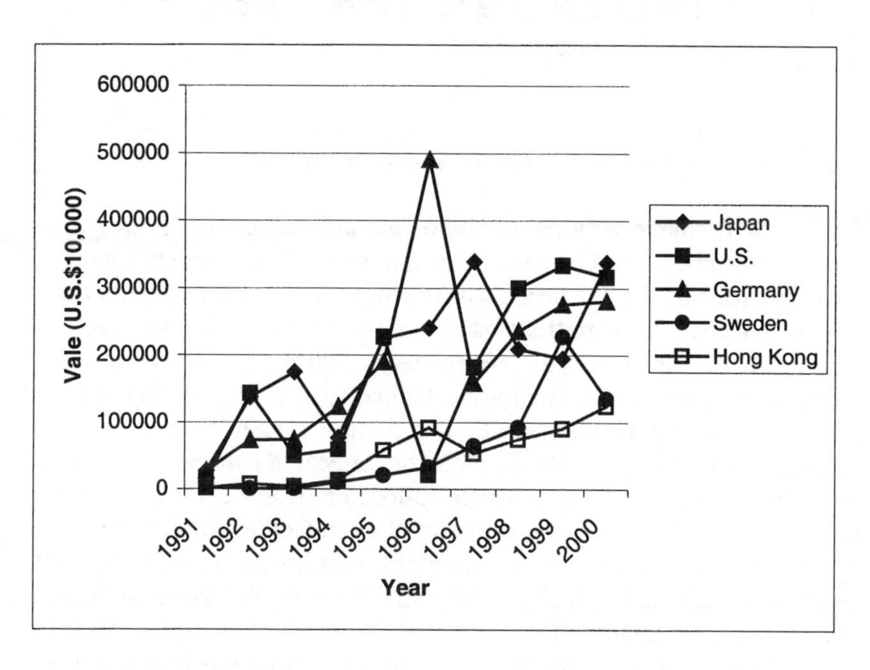

Figure 3.1 Technology import contractual value by country

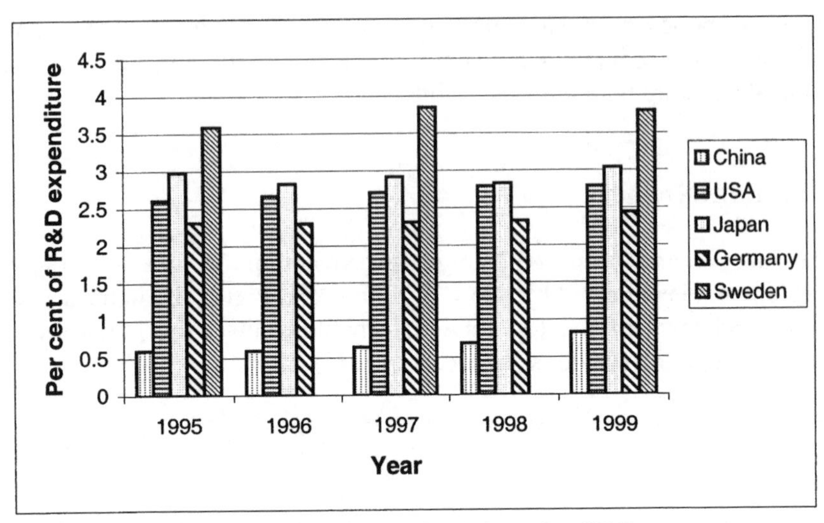

Figure 3.2 R&D expenditure and share in GDP

With the exception of 1996, an anomalous year, investment amounts climbed steadily, indicating a consistent policy and interest in such investments. By the turn of the century, the top three investors were closely bunched, as were the next tier of two countries (China Statistical Yearbook on Science and Technology, 2001). Some of the most strategically important technology in China, such as the development and production of the world's smallest and most powerful microchips, is coming from companies like Semiconductor Manufacturing International and Grace Semiconductor, which re-located from Taiwan. Three of China's most advanced integrated circuit manufacturers are WFOEs; the other five are joint ventures. China plans to construct ten more by 2005 (McMillion, 2002). Ironically, these plants have located large facilities in Shanghai's Zhangjiang High Tech Park that originally styled itself the national Bio-Pharmaceutical base. Due to intellectual property issues, the large (20 out of the world's top 25) pharmaceutical firms located in China sell only drugs available elsewhere, or use China as the site for large scale mandated stage 3 (human) drug trials.

Figure 3.2 indicates the relative amount of gross domestic product in each country invested in research and development, showing China to be significantly less than in the other developed countries.

China's percentage of R&D investment is clearly much lower than that in the more developed country. While Sweden as a smaller country can afford to invest a higher percentage in R&D, as is the case with Japan even during a decade of recession, the U.S. and Germany with large and diverse economies still invest a significant percentage in cutting edge development. The demands of development across the board seem to be holding down China's innovation investments, possibly impacted by major infrastructure projects in lagging regions such as the Three Gorges Dam and Tibetan railway facilities. The importance of learning from more developed countries, through proximity of businesses in park development zones or by other means such as lax intellectual property protection enforcement, seems to continue in the face of relatively small investment in domestic research and development. The lack of government investment in research institutions underlines the emergence of an emphasis on self-funding, pushing many Chinese research facilities to seek out partnerships with foreign firms in order to sustain themselves.

A more upbeat picture emerges with the addition of more R&D in-house facilities by major foreign invested companies in China's STIPs. Big names include Microsoft ($80 million for a research institute in Beijing, US$50 million for an Asian Technology Center in Shanghai), Motorola (US$200 million in an electronics investment base, 650 researchers), GM, GE, JVC, Lucent-Bell, Samsung, Nortel, IBM, Intel, Du Pont, P&G, Ericsson, Nokia, Panasonic, Mitsubishi, AT&T and Siemens (UNCTAD, 2001). Motorola, the largest FDI in China, recently announced plans to invest US$10 billion in China and utilize US$10 billion in Chinese outsourcing by 2006. Major impetus for this approach includes the expected attempt to modify existing product lines to fit Chinese domestic market needs, and the availability of highly trained, low wage technology workers eager for better compensated employment in foreign companies. China's progress in supplying this attractive contingent of high skilled labor has advanced it to a higher level in the eyes of TNCs engaged in the global commodity chain (Gereffi, et al, 1994; Lin, 2000). An interesting study recently demonstrated that Japan is the country most likely to shift production out of other manufacturing sites to China, with 99 per cent of all companies shifting to China (Bronfenbrenner, 2002). This probably reflects several factors, such as the economic pressure for low cost production given Japan's decade of economic slowdown, the close proximity of China, and long-standing Japanese footloose competitive practices.

Asian countries such as Korea and Taiwan overwhelmingly shifted companies from their home headquarters directly to China, while low cost production sites such as Hong Kong largely experienced shifts of companies that were originally headquartered elsewhere. Singapore's shift was 60 per cent natives from that country.

Domestic foreign invested enterprises currently constitute the main source of China's high technology exports, as indicated in Table 3.6. A caution as to these figures: exports from Chinese non-state owned enterprises (SOEs) is not provided, and excluded from the percentages. FDI flows have shifted from the usual low-tech base in footwear, travel goods, toys, bicycles and small household electrical appliances to more upscale products such as computers, electronics, telecommunications, pharmaceuticals, petrochemicals and power-generating equipment. In combination with the presence of 4/5ths of the *Fortune 500* major international companies with manufacturing facilities in China, it is clear that China's STIPs are a major destination for foreign high technology production globally (UNCTAD, 2001). A prime demand of TNCs located in China (and throughout Asia) is increasing the degree to which these parks serve their intended function as a conduit for information flow between government authorities and park-located companies (Nyaw, 1996).

Table 3.6 Exports of high technology products, 1996-2000

Year	Value (US$million)	%from SOEs	%from FDIs
1996	7,681	39	59
1997	16,310	NA	NA
1998	20,251	25	74
1999	24,704	23	76
2000	37,040	18	81

Source: UNCTAD World Investment Report 2001

Prospectus: Impact of WTO

A key dichotomy exists between the policy featured by development authorities such as government entities – and often cited as the source of the WTO provisions most threatening to the continued viability of zones and theory concerning what science and technology corporate configured spaces really need to thrive. The bedrock enticement of policy rests on taxes – their reduction, avoidance, phasing in or out. The competitive advantage of places where science and technology companies are most productive rests on promoting the exchange of ideas. Numerous corporate interviewees stated that tax abatements were nice, but secondary as location considerations. The availability of affordable and obtainable appropriately skilled labor and sufficient infrastructure (transportation, communication, basic electricity, etc.) were far more weighty factors, none of which are affected by WTO conditions.

WTO prohibitions against tax breaks are being used by the Chinese government as a tool for weeding out competing, non-nationally sanctioned development zones. Local governments in future will have to pay for illegal re-imbursements of taxes to companies in their own zones out of their own pockets (Economist, 2002). Western science and technology based firms in established successful clusters were even willing to pay more for these factors if they could just be in an intellectually open environment: the 'soft', non-commodity location feature. Foreign involvement in China's development policy is but one aspect of the usefulness of STIPs, yet it is a fundamental element that co-evolved with economic acceleration plans since Deng first made his mark as a new post-revolutionary leader in 1978. A summary timeline of subsequent major Chinese policy initiatives is set out in Table 3.7.

Table 3.7 Timeline of STIP related policy developments

Date	Development
1978	Reform and Opening declared by Premier Deng Xiaoping
1983	Open Cities established
1984	Economic and Technology Development Zones
1985	Spark Program; Shenzhen High Tech Park
1986	'863' Plan
1988	Torch Plan; incubators established; first high technology development zone, in Beijing
1991	High Tech Industrial Development Zones
1992	Southern Tour by Premier Deng
1997	'973' Plan

Source: Compiled by author

In the race to comply with policy initiatives and garner income for increasingly self-reliant government levels, a variety of new technology-industrial parks have proliferated across the landscape of China, often competing for businesses and spending scarce funds on duplicative infrastructure improvements. Data on the funds invested in such projects is difficult to obtain, since it is an uncomfortable and wasteful process symptomatic of the devolution of central power. Manufacturing zones as techno- and growth-poles are an established part of Chinese government policy for developing designated areas across the country, spreading growth in controlled urban patterns (Johnston, 1999). Unanticipated consequences aside, promotion of STIPs have overall been a central and largely successful part of China's modernization drive. An additional government program entitled 'Seed Fund for Small Businesses' (similar to the U.S. government's National Science Foundation 'Small Business Innovation Research program) was recently established to provide even stronger incentives for starting businesses than the Torch Program. Up to US$100,000 is available in grants to beneficiaries. Within Shanghai, the financial center of the country, private venture capital companies provide additional (and more risk-taking) funds for start-ups. Government-approved international non-governmental organizations (NGOs) such as the Grameen Bank can also offer small loans at very low interest to entrepreneurs, though the government holds the amount of interest down to such low levels that many find China a less interesting place for non-governmental seed money.

The next set of three chapters details the actions and effects of transnational companies in three STIP development zones in south and central coastal China. As noted in a recent study by He and OhUallachain (2001), both TNC and Chinese high technology companies are drawn to cluster in close proximity. These agglomerations feed off the same islands of supporting factors provided by STIPs, needing the rare political, economic, and technical nurturing infrastructure provided in these government designed privileged zones.

References

Boulton, W. and Kelly, M. (1999), 'Information Technologies in the Development Strategies of Asia', Paper for the International Technology Research Institute.

Bronfenbrenner, K. (2002), 'Impact of U.S.-China Trade Relations on Workers, Wages, and Employment: Pilot Study Report', in U.S.-China Security Review Commission, *Documentary Annex Report to Congress,* Government Press, Washington, D.C., pp. 1-127.

Economist Intelligence Unit Limited (2002), *China Hand*, The Economist.

Feinstein, C. and Howe, C., (eds.) (1997), *Chinese Technology Transfer in the 1990s:* Current Experience, Historical Problems and International Perspectives. Edward Elgar, Cheltenham, UK.

Gereffi, G, Korzeniewicz, M. and Korzeniewicz, R. (eds.) (1994), 'Introduction: Global Commodity Chains', *Commodity Chains and Global Capitalism*, Greenwood Press, Westport, Connecticut.

Gu, S. (April 1996), 'The Emergence of New Technology Enterprises in China: A Study of Endogenous Capability Building Via Restructuring', *Journal of Development Studies*, vol. 32, pp. 475-505.

Gu, S. (1999), *China's Industrial Technology: Market Reform and Organizational Changes*, Routledge, New York.

He, C. and OhUallachain, B. (2001), 'Entry Mode and Location: Foreign Manufacturers in Transitional China', Paper presented at Southeastern Division of the Association of American Geographers Conference, Lexington, KY.

Johnston, M. (1999), ' Beyond Regional Analysis: Manufacturing Zones, Urban Employment, and Spatial Inequality in China', *The China Quarterly*, vol. 157, pp. 1-17.

Lin, G. (2000), 'State, Capital, and Space in China in an Age of Volatile Globalization', *Environment and Planning A*, vol. 32, pp. 455-71.

McMillion, C. (2002), 'China's Very Rapid Technological, Industrial and Economic Emergence', in U.S.-China Security Review Commission, *Documentary Annex Report to Congress,* Government Press, Washington, D.C., pp.1-22.

Naughton, B. (1988), 'The Third Front: Defense Industrialization in the Chinese Interior', *The China Quarterly*, vol. 115, pp. 351-386.

Naughton, B., (eds.), (1997), *The China Circle: Economics and Electronics in the PRC, Taiwan and Hong Kong*. Brookings Institution Press, Washington, D.C.

Nyaw, M. (1996), 'Investment environment perceptions of overseas investors of foreign-funded industrial firms', in Yung, Y. and Sung, Y. *Shanghai: Transformation and Modernization Under China's Open Policy*, Chinese University Press, Hong Kong, pp. 250-72.

Pillsbury, M. (2002), 'China's Research Institutes', in U.S.-China Security Review Commission, *Documentary Annex Report to Congress,* Government Press, Washington, D.C.

Ravenhill, J. (1998), 'The Regionalization of Production and Competitiveness in East Asia', in Anderson, R., Cohn, T., Day, C., Howlett, M. and Murray, C. (eds.), *Innovation Systems in a Global Context*, McGill-Queen's University Press, Montreal, pp.174-93.

Simon, D. and Goldman, M., (eds.) (1989), *Science and Technology in Post-Mao China*, The Council on East Asian Studies/ Harvard University, Harvard University Press, Cambridge, MA.

United Nations Committee on Trade and Development (UNCTAD) (2001), *World Investment Report 2001: Promoting Linkages*, United Nations Publications, New York.

Wang, J.C. (1999), 'In Search Of Innovativeness: The Case Of Zhongguancun', in E. Malecki and P. Oinas (eds.), *Making Connections: Technological Learning and Regional Economic Change*, Ashgate, Aldershot, UK.

Wang, J.C (2000), *Chinese Industrial Clusters*, Beijing University Press, Beijing.

Wang, J.C. (2002), 'High and New Technology Industrial Development Zones', in Webber, C. M., Wang, M. and Zhu, Y. (eds.), *China's Transition to a Global Economy*, Palgrave Macmillan Global Academic Publishing.

Wang, J.C. and Wang, J.X. (1998), 'An Analysis of New-Tech Agglomeration in Beijing: A New Industrial District in the Making?', *Environment and Planning A.*, vol. 30, pp. 681-701.

Wang, S., Wu, Y., and Li, Y. (December 1998), 'Development of Technopoles in China', *Asia Pacific Viewpoint,* vol.39, pp. 281-301.

Wu, W. (1999), *Pioneering Economic Reform in China's Special Economic Zones,* The Book Company, Ashgate, Aldershot, UK.

www.chinagate.com.cn/english/559.htm (2002), 'Science Parks', China Internet Information Center.

www.chinatorch.com/stipark/english/page91.htm (2000), 'Shenzhen Science & Technology Industrial Park'.

Yang, C. (August 2000), 'The Historic Progress: The Analysis of Key Economic Indexes and Structures in Shenzhen in the Last Two Decades', Government Press, Shenzhen.

Yeh, A. (1996), 'Pudong: Remaking Shanghai as a World City', in Yung, Y. and Sung, Y., *Shanghai: Transformation and Modernization Under China's Open Policy,* Chinese University Press, Hong Kong, pp. 273-298.

Yu, Q. Y. (1999), *The Implementation of China's Science and Technology Policy,* Quorum Books, Westport, Connecticut.

Yusuf, Y. and Wu, W. (1997), *The Dynamics of Urban Growth in Three Chinese Cities,* Oxford University Press, New York.

Zou, J. (12/18/2000), 'Does China Need So Many Silicon Valleys?' *Beijing Review*, pp. 25-27.

Chapter 4

Multinational Development Zone: Shenzhen

Pioneering a New Municipal Model

In recognition of Shenzhen's importance as a pioneer city for attracting early foreign investment to China following proclamation of the 'reform and opening' movement in 1978, and its science and technology industrial park (designed as a national model for domestically generated economic dynamism), examination of this area is split into two chapters. This first highlights Shenzhen's attraction for foreign companies, while Chapter 9 examines its role as a leading site for domestic technology commodification. Sections in each of those chapters detail the areas of theoretical interest outlined in Chapter 2:

- Characteristics that make a particular place attractive to a science and technology park-based economic cluster
- The types of firms attracted to that industrial district, and case studies of some of these firms or prominent actors
- How these firms in turn gave each place a particular character or niche in a domestic and/or global production chain
- The types of networks constructed in that place, and whether they indicate a 'learning district' or 'satellite branch' or some other mode of interaction
- The role of this location in integrating China into a global economic system

The separate treatment of Shenzhen as the site of both a 'multinational development zone' (Chapter 3) and a 'local innovation learning zone' (Chapter 9) presage the finding that two tracks of development are occurring in China, which occasionally intersect in space but often operate with important distinctions.

In July 1985 the municipal government and the Chinese Academy of Sciences founded the Shenzhen Science and Technology

Industrial Park. Four years later the municipal government established Shenzhen Hi Tech Industrial Park (SHIP), subsuming the original industrial park into one of 27 State science and technology industrial parks (Figure 4.1) (www.chinatorch.com 1999). SHIPs creation marked the birth of the first district set up to combine research from institutions affiliated with the Chinese Academy of Science and foreign company technology transfers (Gu, 1996).

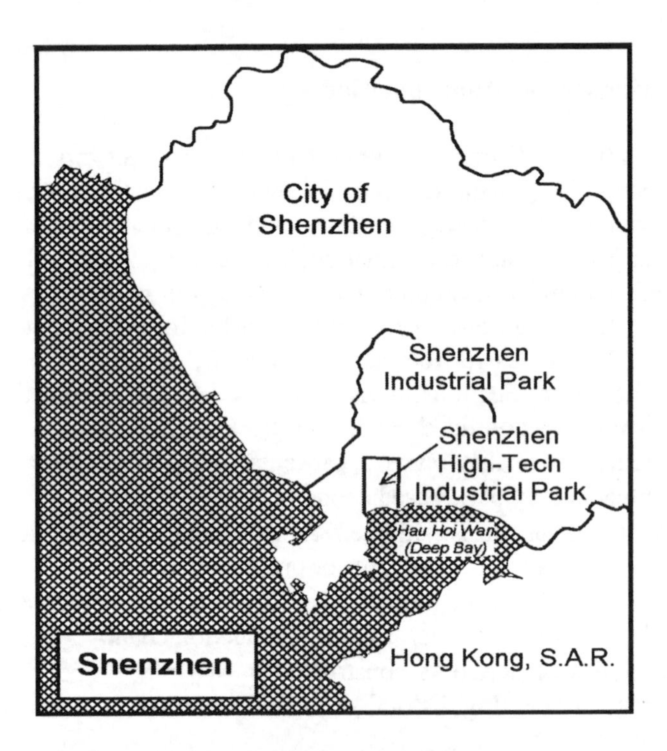

Figure 4.1 Map of Shenzhen's development zones

Investments in Shenzhen in the early 1980s from electrical industries headquartered in Hong Kong signaled the first multinational company activity permitted in China (CSTD, 2000). Scientists and engineers formerly working in the microelectronics field came to Shenzhen to develop other high technology applied products in the new city rising across the river from Hong Kong. Infusions of foreign capital, often funneled through Shenzhen's British colonial city-sister, sustained the early development of this sector in China. By 1997, Epson and Compaq of Shenzhen were ranked #7 and #8 of the top ten foreign enterprises in China earning much needed foreign exchange through export.

However, the dominance of Hong Kong enterprises and the type of activity they conducted in Shenzhen indicated that primarily (and predictably) low-technology intensive operations relying on low cost labor for assembly of parts with high technology embedded components from more advanced countries were to be Shenzhen's fate. Advocates of 'appropriate technology' argue that this is entirely suitable, considering the level of available resources (Wu, 1999). However, the government of China just as clearly desires to accelerate the acquisition and utilization of technology by every means available. They put in place policies such as frameworks for intellectual property protection (with woefully lax enforcement), joint ventures under various arrangements (equity or contractual), nationalization of company management (to accelerate the training of Chinese), and tax advantages in order to attract foreign companies to favored sites. To this was added the particular practices and policies of local park and zone managers. Shenzhen stands out as one, which carefully seeks to follow Hong Kong's demonstrably successful model of minimum regulations and maximum speed in processing, as well as policy transparency, and ease of access to Hong Kong's well developed shipping facilities. The results of these efforts remain mixed, and are addressed throughout the two chapters set aside for examining Shenzhen's experiences.

Companies with non-Chinese headquarters operated in China provided they were technologically advanced, held promise for generating foreign currency through export, and filled an area of national need (Weidenbaum and Hughes 1996). These requirements have been relaxed in the last several years, and will be largely outdated by World Trade Organization provisions that aim for open and equal treatment between local and multinational companies. By the late 1990s, joint ventures constituted 64 per cent of Shenzhen's high technology enterprises, accounting in 1996 for 70 per cent of Shenzhen's total product value. This form of organization remains a tricky strategy, most useful at the time of a company's initial entry to the Chinese business environment (Lan, 1996).

Shenzhen Hi Tech Industrial Park: Proving a Point

SHIP inhabits a relatively small section (11.5 sq. km.) within the larger and better-known 327.5 sq. km industrial park. The only importance of the name distinction is as a geographic locator within the original industrial district. The central part of SHIP is the older core area. Most of the multinational corporations are located in the northern section, since it opened in the early 1980s when they first came to China and sought location sites (Figure 4.2). While comprising only 4.5 per cent of the total companies located in SHIP, multinationals principally in IT and optical-electronics include major global players such as IBM (U.S.), Olympus (Japan), Lucent (U.S.), Harris (U.S.), Compaq (Japan), Epson (Japan), Philips (Holland) and Thomson (France). As an extra bonus, the large transnational corporations IBM, Compaq, and Epson brought other supplier companies with them to continue relationships established in other global production sites. The top ten multinational corporations in China include three from Shenzhen; 15 firms fall in *Fortune* magazine's list of the top 'Global 500' (CSB 2000a). Park developments advance the intent behind the establishment of the Shenzhen: to show the outside world that China was open for business.

Within the boundaries of SHIP, designers attempted to delineate separate but integrated areas for research facilities, manufacturing plants, residential quarters of various cost scales, education (schools for children of employees residing in the area and higher education institutions), incubators, and large units with food services within walking or biking distance. Reflecting the prevalent trends at the time of construction, newer areas preserve more designated and designed greenspace. Aside from most of the foreign companies, the older northern section contains lower cost worker dormitories in the usual rows of simple cement blocks. Surrounding companies often rent these for convenient employee housing. The middle district contains SHIP development headquarters, a cancer hospital, several joint venture companies with ties to Peking University, and a number of establishments providing services such as food for workers and hotels for visitors. The southern district contains the most green space; it is still under construction as a large lake formerly on the property needed to be drained. The joint university park area discussed more fully in Chapter 9 will occupy space in this newer area. The prevalence of Chinese companies and research entities in this section reflects the evolution through time and space toward more domestic headquartered R&D businesses in Shenzhen.

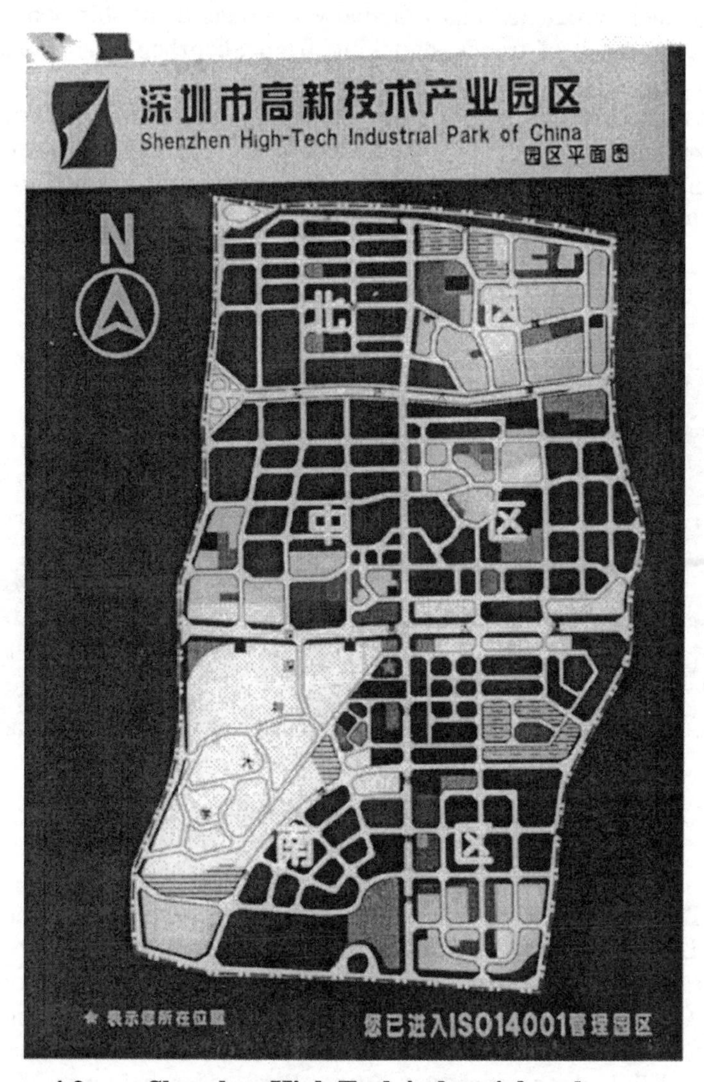

Figure 4.2 Shenzhen High Tech industrial park

Though over 90 per cent of the companies in Shenzhen report exporting their products, many actually re-route their shipments, as evidenced by lines of trucks stretching from Shenzhen in and out of Hong Kong, looping back across the border to sell their cargo in the domestic market (Figure 4.3). In this way, many companies (particularly FDI firms who are strongly encouraged to export) can show export figures that actually mask how much of their product is destined for the Chinese market.

Figure 4.3 Container cargo leaving on the overland loop

The value of Shenzhen's high technology products has steadily increased, as detailed in Table 4.1, and is heavily oriented to export. The extent to which these apparent export products are actually re-exported, entering China from Hong Kong to their final destination in the Chinese domestic market, is difficult, if not designed to be impossible, to determine. While the practice contravenes legal codes, evidence of such activity is abundant – and potentially highly profitable. Computer and software products constitute the greatest amount by far of Shenzhen's products; with telecommunication equipment, they represent more than 70 per cent of export value since 1996. This reflects the predominance of this sector in the Pearl River delta region as a whole.

Table 4.1 Measures of Shenzhen's high technology production

Category	1992	1993	1994	1995	1996	1997	1998
HT value	47.3	77.3	146.2	225.8	347.6	474.5	655.2
Electronic information	38.9	66.6	128	211	299.3	432.2	602.1
Mechanical/ electrical	5.1	4.8	6.7	4.3	30.3	20	26
New materials	2.4	4.6	8.4	6.6	8.8	13	9.6
Biotechnology	.8	1.3	2.8	3.7	7.8	9.2	17
% Annual increase	107	63.4	89	54	54	36	38
% High Tech of Shenzhen gross industrial value	12	14	17	21	29	35	35
High Tech exports	2	3.5	10	17	28	37	44
% Annual increase	82	82	72	66	30	19	
Total exports	120	142	183	205	212	254	264

Source: China Science and Technology Department (2000)

Multinational companies heavily rely on government intermediaries for local information and on their headquarters for funds. The national Foreign Enterprise Association in Shenzhen informs multinational corporations of new directives that affect them and in turn solicits questions and advice. The government also provides tax reduction (up to 50 per cent) for high technology companies. The government also provides an initial two years of low taxes, followed by three more years of tax reduction (up to half) for high technology companies.

As the historically first place in China opened to foreign business in both the 19th and 20th century, the Pearl River delta took its attractiveness for granted. As in the early 1900s, the dawn of the 21st century finds this region competing again with the Yangtze delta, pitting metropolitan Guangzhou against metropolitan Shanghai in selling their business-friendly environments to companies from more advanced countries. The prize for the Chinese involves jobs and capital – fiscal, physical, and intellectual potential knowledge transfers. As the central government poured funds into the South in the 1980s, the 1990s has been the decade of redress for coastal favoritism inequities, fine-tuning the regional balance in favor of long-neglected Shanghai (Fan, 1997).

Guangzhou finds itself with a relatively dilapidated airport and shipping facilities, as well as a rundown urban setting compared to Shanghai's shiny showcase in Pudong. Taiwanese investments, 70 per cent of which flooded into Guangdong province and Fujian province to its north (directly across the Taiwan Strait with its own Special Economic Zone in Xiamen designed especially to attract such investments) in the 1980s, continue to migrate to Shanghai and neighboring Suzhou (Landler, 2001) (see Chapters 6 and 7).

While Shanghai boosters stress its financial assets, with special prerogatives permitting foreign and Chinese institutions to conduct operations there as in the 1930s, and ties between Shanghai and current Beijing leaders who frequently apprenticed as mayors of that city, Guangzhou province centers such as Shenzhen tout a few unique assets of their own. The enterprising independence of the Southerners (in matters such as speaking and broadcasting in their own Cantonese dialect rather than the northern Mandarin dialect designated since the 1920s as the national tongue) appalled Deng Xiaoping in his 'Southern tour' of 1992 and led directly to the central government's redirection of funds into central coastal Shanghai. This same independence is attractive to businesses that find Shanghai overly bureaucratic and slow compared to the more 'flexible' Southerners. As a balance to the resurgent central region, Beijing maintains political and personal ties to Shenzhen with economic coattails, as discussed in this chapter and Chapter 9, which deals more with domestic university-related high technology ventures.

Delta business and political leaders have also reached some compromises with their central China rivals, seeking to find identifiable production niches to blunt the competition for scarce foreign high technology giants willing to come to China. In computer assembly and manufacturing, for example, Shanghai attracts primarily semiconductor and notebook computer companies, while the Guangzhou-Dongguan-Shenzhen region retains consumer electronics and desktop computer specialties. Giants such as Motorola, with a major plant in Tianjin outside Beijing and a factory in Pudong, eye Guangdong as a regional expansion as well (Landler, 2001). This practice of carving out a spatially identifiable separation of economic activity extends to cities within the Pearl River delta as well. More peripheral cities such as Dongguan attract a high percentage of foreign investment in manufacturing (@70 per cent), while larger entities like Guangzhou and the Special Economic Zone of Shenzhen demonstrate a much lower percentage of FDI in the secondary sector. Overall, attraction of foreign invested companies drove urbanization in these formerly heavily agricultural areas in the previous two decades (Sit and Yang, 1998; Lin, 2001). While China has been the largest recipient of FDI since 1993, and the U.S. the largest importer of Chinese products, China surpassed the U.S. as the biggest trade partner of Hong Kong since 1985, a tribute to the Pearl River delta's geographically proximate strength. Both within 100km of Hong Kong, High Tech Shenzhen and its lower-tech neighbor Dongguan accounted for almost 40 per cent of all FDI in the Pearl River Delta in the early 1980-1993 time period, with approximately the same share of PRD-based Hong Kong investment in the two cities (Sit and Yang, 1998).

Firms with a Global Reach

Two firms with foreign ownership are profiled below, to provide a more detailed look at the motivation and challenges for companies coming to China. The first functions as part of a global conglomerate, headquartered in Canada and Melbourne, Florida. Local management comes from Hong Kong and South China natives. The second company, while considered 'foreign', is owned and managed by a returned Chinese graduate of an American university. While not a native of Shenzhen, and despite his marital family still residing in the U.S., he pioneers his innovative medical device manufacturing venture within SHIP.

The Harris Company

The U.S.-Canadian Harris Company, which supplies medium and low wave digital microwave systems, telephone and data transmission capacity throughout China, is a SHIP example of large transnational company that typically entered China through an initial joint venture (Harris, 1999). Harris started in Shenzhen in 1993 as a joint venture with a Chinese partner in order to be quickly ingratiated into a connective 'guanxi' network. However, management differences and a desire to leave congested downtown Shenzhen led to the partners' split and Harris' subsequent removal to cleaner, quieter surroundings in SHIP. Company officials interviewed indicated that a major consideration for moving to the High Tech park lay in its well-known reputation as a prestigious address with fair management practices. Since its re-location in SHIP, Harris grew from 20 to over 120 employees.

Most supplies and technical parts for products made in Shenzhen originate in North America. Some local suppliers are used for simple products such as cardboard packing material, despite the common complaint that the search for local sources often founders for lack of sufficient quality standards. The government provides an initial two years of low taxes, followed by three more years of tax reduction at a rate of 50 per cent for high technology companies. Harris' lobby sports several prominent plaques attesting to its status as a tax compliant corporation. The company prides itself on following local requirements in all its various global locations. North American companies have the reputation of being unusually tax compliant. They reputedly have found this to be a practical matter, as it provides accounting predictability in budgeting and some protection against sudden and unexpected demands for large sums of unpaid back taxes. Harris is a member of the Foreign Enterprise Association set up by the government to inform foreign companies of the latest regulations and explain enforcement procedures and ramifications.

A tour of the shop floor at Harris revealed a typical gender divide. Technical workers were overwhelmingly male, while females held assembly line positions. All were young (20 somethings). Chinese universities in other cities provided 95 per cent of the college educated employees, or 85 per cent of the total college-educated employees. Most recruits came from inner China and other companies, not straight from universities. Some organizations link electronic and engineering workers nationally, but most function through the Harris company network. Most technology transfer occurs within the company through training nationally recruited workers, which is the main connection to China. Higher-level technology comes from the Canadian company headquarters. A small number of selected Chinese workers are occasionally sent to Canada for advanced training and orientation to Harris corporate procedures, but this carries risks. In the past some of these employees have disappeared into Canada, or later quit Harris and immigrated to Canada. On the whole, the management felt that doing business in China, and in SHIP, was a quite profitable and worthwhile venture. All top executives were ethnic Chinese; some came from Hong Kong, others were Pearl River delta natives.

Mindit Corporation

The Mindit Corporation represents another type of development path for companies located in China. A manufacturer of magnetic resonance imaging machines, Mindit was started in 1998 by a group of Chinese students returning to China following graduation from various American universities. These alumni maintain ties with former professors; one is featured (in the photographic role of 'Western expert') in their corporate brochure (Mindit, 2000).

When queried as to why the owner of Mindit came to Guangdong province rather than returning to his native city further inland to start his business, the founder cited several advantages as well as drawbacks to his current location. Shenzhen's perception as a hotbed of computer-related High Tech and export center draws Chinese companies, as well as the predominance of small and medium Chinese firms, compared to multinationals with a base in Shanghai or Beijing. A major drawback remains the lack of a significant university or research center compared to other major cities. Potential employees at High Tech firms prefer to stay close to their alma mater and home provinces, including major metropolitan centers in inner China such as Wuhan, Chengdu, and Xi'an. Many of Mindit's employees are therefore recruited from other companies in SHIP, as well as through online and newspaper ads.

Technology transfer infusions come from personal ties to university student exchanges with Chinese universities, and consultations with local professors. Local operations still centered on manufacturing, with some modifications in design, while the R&D edge remained in the United States. The founder noted that Shenzhen City offers a 'Company Association' and a 'President's Club'. He is personally involved in reorganizing the local returned student's association. Branches of this organization exist throughout China.

Several informants mentioned the high quality of the local K-12 schools, since the Park and city authorities have more power to improve these amenities than to raise the college standards One individual's family not un-typically remained in the U.S., in part because of the difference in the style of Chinese education (memorization) and American style education (broad based, more creative). Lack of a critical mass of a middle class in Chinese society was predicted to remain a disincentive, slowing the return of migrants through the next decade.

Successful business people who left China to attend college abroad after the disillusionment and instability of the Tiananmen Square incident now anticipated increasing migration back to China, such as occurred in Taiwan in the 1980s. Facilities such as those in Hsinchu, the High Technology development park catering to returned students and their families with special Chinese culture and language classes outside the capital Taiwan, were seen as coming on the horizon, in an eventual virtuous cycle over a period of time.

Unlike the situation in Taiwan, however, some of these former students considered that, given the transitioning political and economic situation, it was still premature to return to live and work in the mainland. They preferred instead to retain their lucrative and comfortable positions abroad, investing in or entering into joint ventures or consulting relationships with companies in China. The maturity and vigor of China's financial institutions and capital structure constituted the one of the main stumbling blocks to full participation, though the potential for growth in the domestic market and the lure of heading one's own business remain attractive. Other problems anticipated by these potential returnees were China's underdeveloped marketing and management capabilities. Numerous links nevertheless connect Chinese capital and business ventures in Taiwan, Mainland China, and American technology centers such as Silicon Valley (Gilley, 1999).

Centrally Coordinated Networks

Managers echoed other companies' representatives in the opinion that association with SHIP proved advantageous via a prestige address, well-maintained attractive surroundings, good quality buildings and infrastructure, all at a reasonable rate. Many managers lived nearby in the abundant supply of new apartment complexes. Unlike most areas in China, bikes were few and vans predominated to furnish worker transport. The government and Park authorities provided affordable, close-by apartment facilities for workers, rented by local companies. Government support through infrastructure provision and policies assisting startup companies received positive reviews. Changes in the migration laws (*hukou)* made it easier to supply labor at all skill levels, shifting some of the flow of highly skilled technicians from the more expensive big cities such as Shanghai and Beijing. Government also provides land at easy rates for new companies, and the usual packet of graduated tax relief, depending on the type of business, number of employees, time it came to the area and number of years in that location.

On the whole, Shenzhen has served as less of a technology transfer conduit, rocketing China's upgraded industrial development, than originally hoped. The multinational companies in particular are primarily attracted to Shenzhen by its local low cost labor and easy access to Hong Kong. Within Harris, for example, technology primarily flows from the Canadian corporate side, and top employees go there to train for 4-8 weeks. Most technology transfer takes place through such labor exchanges. Some employees left Harris-Shenzhen for Canada, immigrating there to start new careers in other areas if necessary. On the technology side, one product was developed locally specifically to fit the Chinese market, like Motorola's adaptation of its popular branded communication device to a local Chinese product.

As a district, SHIP appears to be especially well managed, consciously seeking to emulate the successful model of near neighbor Hong Kong in matters such as 'one-stop' expedited form processing. Necessary support amenities such as housing, schools, food, and public bus lines are convenient and run the full scale of affordability. The district's policy of Mandarin dialect advocacy creates an inclusive atmosphere for labor from all regions of China. Shenzhen provides a vital bridge to the developed periphery, the Asian Tigers of Hong Kong, Taiwan, and South Korea (Naughton, 1997), as well as employing a variety of more technology-intensive labor than in other areas of south China. Interestingly, the other Asian Tiger of Singapore chose to locate its business capital interests to the north of the Pearl River delta, in Shanghai and Xi'an, as explored in the following chapters.

References

China Science and Technology Department (2000), 'Report on High Tech Industry Development in Beijing', *Development Report on China's New & High Tech Industry*, China Science Publishing Department, Beijing.

Fan, C. (1997), Uneven development and beyond: regional development theory in post-Mao China, *International Journal of Urban and Regional Research* vol. 21, pp. 620-39.

Gilley, B. (1999), 'Looking Homeward: Tiananmen Exiles Lead Way Into China's Hi Tech Future', *Far Eastern Economic Review*, vol. 162, pp. 50-52.

Harris, (1999), 'Communications Equipment for Voice, Data and Video in the 21st Century, 1999 Summary Annual Report'.

Lan, P. (1996), *Technology Transfer to China Through Foreign Direct Investment*, Ashgate, Aldershot, UK.

Landler, M. (6/23/01), 'China's Once Undisputed Business Gateway Now Has Competition', *The New York Times*, B1-2.

Leung, C.K. (1990), 'Locational Characteristics of Foreign Equity Joint Venture Investment in China, 1979-1985', *Professional Geographer*, vol. 42, pp. 403-421.

Leung, C. K. (1993), 'Personal Contacts, Subcontracting Linkages, and Development in the Hong Kong-Zhujiang Delta Region', *Annals of the Association of American Geographers*, vol. 83, pp. 272-302.

Lin, G.C.S. (2001), 'Metropolitan Development in a Transitional Socialist Economy: Spatial Restructuring in the Pearl River Delta, China', *Urban Studies*, vol. 38, pp.383-406.

Mindit, (2000), Corporate brochure.

Naughton, B., (1997), (eds.), *The China Circle: Economics and Electronics in the PRC, Taiwan and Hong Kong*, Brookings Institution Press, Washington, D.C.

Park, S. O. and Markusen, A. (1995), 'Generalizing New Industrial Districts: A Theoretical Agenda and an Application From a Non-Western Economy', *Environment and Planning A*, vol. 27, pp. 81-104.

Shenzhen Municipal People's Government (1999), 'Provisions for Further Boosting the Development of High-Technology Industries', Government Publishing House, Beijing.

Sit, V.F.S. and Yang, C. (1998), 'Foreign-Investment-Induced Exo-Urbanization in the Pearl River Delta, China', *Urban Studies*, vol. 34, pp. 647-77.

Wei, Y. D. (1999), *Regional Development in China: States, Globalization, and Inequality*, Routledge, London.

Weidenbaum, M. and Hughes, S. (1996), *The Bamboo Network: How Expatriate Chinese Entrepreneurs Are Creating a New Economic Superpower in Asia*, The Free Press, New York.

Wu, W. (1999), *Pioneering Economic Reform in China's Special Economic Zones*, Ashgate, Aldershot, UK.

Chapter 5

Multinational Development Zone: Dongguan

Introduction

Often considered to be the premier example of an industrial cluster built by attracting foreign and nearby Hong Kong business investment to China based primarily on a large pool of hard-working, low cost labor (Tong, 2000), Dongguan provides a picture of the lower end of technology-related clustered activity. Starting with Hong Kong on the eastern shore representing the relatively highest level of innovativeness and pay scale, the Pearl River Delta's string of manufacturing towns next moves west to Shenzhen with its combination of industrial and High Tech parks, the provincial city of Guangzhou with its universities and institutes, and finally Dongguan (Figure 5.1).

Economic clusters within this large township run the technology gamut from textiles – the classic starting point for industrialization around the globe since the beginning of the Industrial Revolution – to assembly operations for globally exported and labeled personal computers (PCs). Apart from its attraction of foreign businesses, many of which are participants in the so-called 'race to the bottom' search for lowest cost sites, Dongguan is also the headquarters for a highly successful township and village enterprise (TVE) form of public-private conglomerate. This chapter explores the TVE mode of clustered companies, and their role in creating a location where foreign business forms are adapted as well as attracted across the pay and technology scale. The 'Winnerway' corporation, hailed as one of China's best known and most successful township and village enterprises, is profiled for insights into this transitional (and transitioning) type of organization and relation to a Chinese type of industry cluster and technology park.

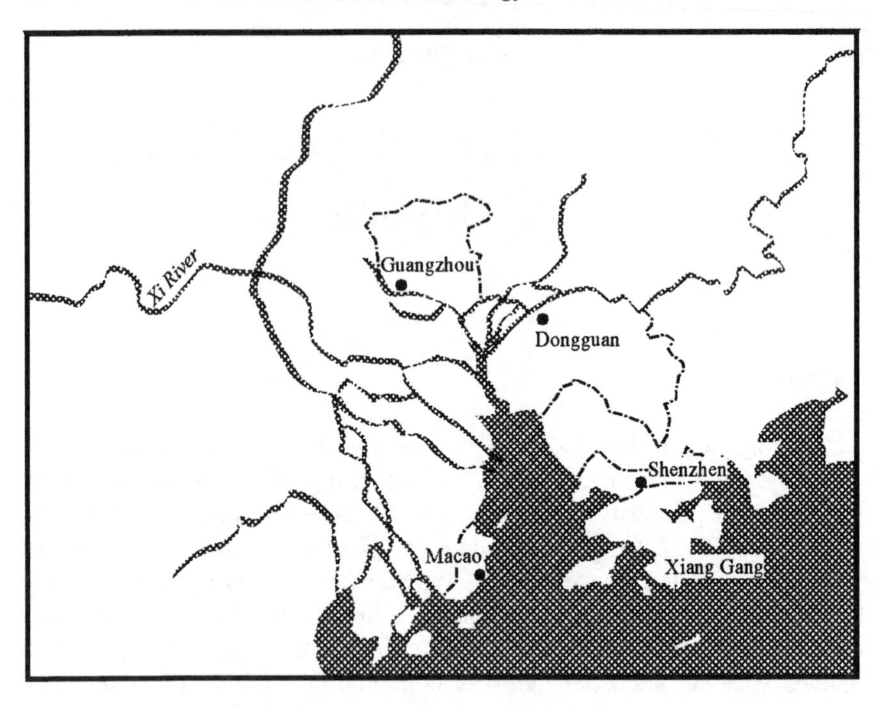

Figure 5.1 Dongguan's location within the Pearl River Delta

Hong Kong can be considered a type of 'transferred industrialization', integrating Western free market characteristics in a Chinese setting. Businesses draw on a vast pool of low cost, hard working labor (largely from mainland migrants) linked to global export routes by an excellent transportation infrastructure. Hong Kong's deep-water port is still the best in China, and its second international airport was a departing gift from the British colonial administration. Since the 'reform and opening' period launched in 1978, these locational advantages spilled across the former border to include the Pearl River delta region in 'an extended, cross-border industrial economy' that intensified following Hong Kong's reunification with China in 1998 (Sit, 1998, p. 881).

A cooperative form of business known as 'out-processing' characterizes relationships between companies who are based in China and provide the labor, land and plant, and infrastructure in exchange for the foreign input of capital, design and marketing of the product. Within the different forms of joint ventures, some businesses divide the profit in proportion to the value invested by each side (an equity joint venture), while others divide based on a pre-arranged formula (a contractual joint venture). The chief draw for businesses coming to this area indisputably lies in the abundance of low cost labor from throughout China, including Western provinces such as Sichuan. Another form of cooperative arrangement lies in 're-processing', with the Pearl River delta supplying manufacturing and assembly capabilities for Hong Kong's other features. The manufacturing aspect of this arrangement between Dongguan enterprises and services supplied by partners in more advanced global regions extends well beyond transactions involving Hong Kong, but often flows across Hong Kong's borders and shipping facilities.

This contemporary economic interconnectedness of regions marks a reversal of Mao era policy (1949-1977) wherein 'cities develop industries and villages develop agriculture' (Ma and Fan, 1994, p.1625). Instead, the appearance of small to large scale manufacturing activities in the former countryside has led to the burgeoning of towns and an 'urbanization from below' that results from indigenous economic initiative as the central government withdraws its formerly all-pervasive directive intervention. Now leading local politicians are also leading shareholders of local Township and Village Enterprises (TVEs), whose stock is widely held but managed by a small committee. The spatial redistribution of growth areas is occurring not only among coastal large cities, from south to north, central to inner regional powerhouses, but on the semi-periphery between middle-size cities in a region connected by transportation linkages and access to a distinguishing resource base. Dongguan illustrates the low-tech assembly of High Tech products, and a dual track of transnational corporation and indigenous organizational forms. China is indeed integrating into a global capitalist economy, but with its own types of connections that are being invented as the state disengages politically to create more economic space (Lin, 1999).

The Pearl River delta region where Dongguan is located covers 47,430 sq. km, with a population of over 20 million. This results from a highly agriculturally fertile climate and soil for rice and vegetable production with a historically dense and rapidly increasing population. These factors plus access to Hong Kong's harbor made Dongguan a major sending area for Chinese emigrants. Now it receives investment money from descendants of these emigrants, and new immigrants from poorer inland China. Although migration restrictions have been eased since 1978, the intent is still to control more tightly the growth of population in large cities in favor of growth in the intermediate size and geographically intervening townships.

The peri-urban zone that developed in the delta represents a functionally interdependent mixture of agricultural and industrial activities (Lin, 2001a). In China, this intermediate zone can extend from 150-300 kilometers outside the built-up borders of a city – borders which can still be sharply visually delineated by the sudden appearance of a ringing wall of tall apartments housing urban residents. The larger zone and density occurs more frequently in coastal Chinese cities than in large inner China urban areas (Webster, 2002). This type of footloose economically driven migration – some legal, others without prior permission, relatively poor, young, and in close proximity to more prosperous areas – has also led to more crime and exploitation than in the preceding five decades. Large elements of a 'floating population' often active in the day labor construction arena, are frequently seen around booming STIP and other industrial district sites which mushroom on land along the more affordable and available urban outskirts. The phenomenon occurs throughout large Chinese cities, however; this is a country, after all, with over 53 cities having a registered population in excess of one million.

Dongguan setting

This large area (2,465 sq. km.) Pearl River delta city lies almost equidistant between Shenzhen and Guangzhou. It's historical fame rested on its output of tropical fruit such as litchis and longans, and migrants leaving its fishing ports for economic advancement overseas. Dongguan doubled its population since the 1980s from a flood of inner China migrants coming to work in factories largely set up by Hong Kong and Taiwanese *fei yang guei zi* (pretend foreigners), in a sense reversing the prior human tide that flowed only outward. These entrepreneurs were attracted by several advantages advertised by Dongguan. These featured the town's ties to overseas Chinese as a traditional sending district for emigrants and good transportation infrastructure connections Beijing and Hong Kong through the Pearl River delta. Most important are the wage rate differentials with production sites in coastal cities, due Dongguan's location between poorer and underemployed inner China and bustling cities with more restrictive resident permit policies to its immediate east such as Guangdong and Shenzhen. By 1998, Dongguan's exports boosted the region into place as the third most export producing city in China (Department of Geography, 2000). Dongguan's performance relative to surrounding Guangdong Province, the leader in manufacturing exports, and China as a whole, is detailed in Table 5.1.

Table 5.1 Dongguan in relation to Guangdong Province and China

Category	Dongguan	%Province	%China
Population (10,000)	149	2.9	.12
GDP (100m RMB)	356	4.5	.48
Primary industry (100m RMB)	31	3.07	.22
Secondary industy (100m RMB)	192	4.8	.49
Tertiary industry (100m RMB)	133	4.54	.51
Manufacturing production value (100mRMB)	732	5.3	.61
Agricultural production value (100mRMB)	51	3.17	.20
'Innovation' fixed assets	327	2.45	.18
Total I&E (100m US$)	233	17.9	7.19
Imports (100m US$)	102	18.78	7.28
Exports (100m US$)	131	17.31	7.13
Utilization, foreign capital (100m US$)	15	9.93	2.56

Source: Adapted from 'Statistical Yearbook of Dongguan' (1999)

Geographic proximity to Hong Kong, and access to the low cost labor supplied by migrants from China's hinterland, account for its current economic position. One fifth of the population purportedly possesses familial connections with the prosperous former British colony and bustling duty-free port city. A busy expressway connects the two Pearl River delta cities, on either end of the production chain. Reportedly over 90 per cent of the former British colony's manufacturing and production work relocated to this area since proclamation of 'opening and reform' in the late 1970s (Department, 2000). The real source of Dongguan's competitive advantage comes from its hardworking, heavily female migrant labor pool. Rows of clean stations and uniformly dressed workers in white blouses and hair scarves are reminiscent of Mexico's *maquilladora* setting serving a similar purpose (Figure 5.1). Dongguan is also known for its textiles, usually the first industry on the industrial road due again to the competitive advantage of low-cost, hard-working, heavily female labor.

Figure 5.2 Female factory workers

A look at high technology production as it developed in the boom years of the 1990s in Dongguan is provided in Table 5.2. Projects deemed most likely to succeed as high technology infused ideas receive Torch sponsorship and are often located in STIPs. The Spark program, set up in the late 1980s, promotes the application of 'small tech' in largely village and township settings, often in 'Spark Tech intensification areas'. The prominence of Spark projects relative to Torch projects indicates the relative degree of rural activities in this area. Rapid growth in the percentage of technical workers and the jump in value of export products indicate the rapid shift in economic activity to the more prosperous relatively technology-infused production. The disparity noted in figures in the table below might be accounted for if Dongguan is not considered included in the total numbers for Guangdong Province (in which it is, in fact, a part).

Table 5.2 Development of High Tech in Dongguan

Category	Units	1995	1998
# Hi-tech companies		11	39
Production value	10,000 RMB	305,520	1,137,000
Production value of hitech products	10,000 RMB	314,304*	926,000
Variety of products		27	196
Export revenues	10,000 US$	5,422	59,602
# employees		6,400	22,053
# technical personnel		832	5,427
% workforce technical		13%	25%
Torch projects '95-'98			
Guangdong		1	4
Dongguan		6*	1
Spark projects '95-'98			
Guangdong		5	6
Dongguan		10*	3

*Note: apparent contradiction, not unusual in Chinese statistical sources
Source: Adapted from 'Statistical Yearbook of Dongguan' (1999)

Dongguan's economic activities generally fit Park and Markusen's (1995) model of a 'satellite industrial platform' characterized by inter-district mobility of labor, non-local linkages and non-local networks. Within the personal computer (PC) sector, Dongguan functions as an export processing center in a global network with strong connection to overseas Chinese through Hong Kong, but weak local linkages to other production chain activities. A Taiwanese-based PC network flourishes due to the cost advantage of inexpensive labor and land, an adequate transportation infrastructure, and non-local (foreign) management. The manufacturing process is integrated between many small and medium firms operating through specialization, collaboration, and social linkages. Products also feed a growing domestic market for PCs in China, at the lower end of the price chain. Locally produced intermediate products hold down costs.

Over 30,000 Taiwanese live and do business in the Dongguan area, drawn there by its more developed infrastructure relative to

Xiamen, the Special Economic Zone set up in Fujian province that is geographically closer to Taiwan. They are particularly prominent in personal computer (PC) manufacturing, a leading high tech product of this city. In 1998, PC production represented almost half of the township's gross industrial production value. Official statistics put the number of Taiwanese invested firms in mainland China at 45,000 by the end of the 1990s, one-third in Guangdong province and a quarter of these in Dongguan. Despite unfavorable publicity stemming from illicit activities engaged in by Taiwanese businessmen, the township also features a Taiwanese elementary school and other civic networks for more settled residents with cross-Strait ties (Department, 2000).

Migrant workers more than doubled from 1990 to 1998; the large number of presumed illegal immigrants makes it difficult to use precise figures. This situation particularly applies in some of the 30 Dongguan towns and villages where small and medium factories located. They are often single, but not necessarily unmarried. In many cases families are left in the countryside hometowns while the wage earner works at a higher paying job than locally available and sends back as much as can be saved to support those left behind. The boom of manufacturing employment in this formerly rural agricultural region functions as a crucial mechanism for absorbing underemployed labor in both the local and interior regions. It also functions to anchor this labor in areas peripheral to the largest cities, such as Guangdong and Shenzhen, or at least retard their migration by providing a closer employment venue (Fan, 1996; Lin, 2001). In one of the colorful phrases used to describe the actions of government planners to attract investment, one official declared that they 'planted grass to raise the cattle', in order to produce 'beneficial results to milk' (interview, 2000).

Township and Village Enterprises

Township and village enterprises constitute a unique Chinese form of local group entrepreneurship. They arose in the 1980s in response to the de-collectivization of agriculture, which freed a number of under-employed villagers for other occupations. At the same time, the central government passed on responsibility to the local level to manage the new labor available, and permitted local government to keep a greater share of locally generated funds as an incentive for new types of income (and tax) generation. The new mandate became self-funding, if the opportunity was successfully seized. Rather than passing into a private enterprise system, this encouraged the development of a local state corporatism (Oi, 1999).

The ownership and management of TVEs at present lies in the gray area between a cooperative village-based framework and clearly delineated property rights. They function in the twilight zone of 'market socialism' transition, between the collapsing of giant state owned enterprises (SOEs) and the rise of wholly private enterprises. Their marginal and unclear nature has led one scholar to declare this a prime argument for the need for institutionalization, or clarity and a fixed order, particularly in regard to rural area legal statutes (Williams, 2002). Investment in TVEs slowed since the late 1990s, causing concern that they might not provide the answer to unemployment and an innovation dearth in rural areas. This sector of China's vast economy employed almost a quarter of the workforce, accounted for over 40 per cent of industrial output, and fell by almost ten per cent in 1995 (Smith, 1997). Their numbers continued to shrink through the early years of the 2000s. Areas of difficulty revealed China's underlying lack of business management experience and expertise. Over-investment was linked more to what seemed like a good idea rather than market research or response. Loosely defined property rights contributed to lax performance lacking profit motive. Unsophisticated technology and poorly trained technicians and personnel produced low quality goods for non-existent or already crowded markets. Many also went for lowest cost production methods and ended up heavily polluting their environment. Fraud and local level corruption afflict TVEs and much of the rest of the economy. A few successful examples flourish in places particularly suitable for this type of venture, however. The following section profiles one of the largest and most interesting, from the Pearl River Delta.

Case Study: Winnerway

Dongguan's former peasants drew on two sources of geographic strength in founding Dongguan Winnerway Industrial Zone Ltd. Company in 1987. One was their location just to the west of the rapidly developing city of Shenzhen, and the other was their historic ties as a sending district for emigrants throughout the world, particularly through descendants of residents now living in southeast Asia and the United States. Dongguan thrives as a particularly successful example of township and village enterprises in the rich agricultural and business environment of Guangdong Province.

Utilizing an ownership structure unique to China, the Winnerway Corporation began in the late 1980s as a township enterprise venture. The State government had recently authorized this new type of organization in the hope of generating grass roots entrepreneurship, spreading the newly generating wealth and opportunities from creating enterprises to the countryside, and attaching peasants to land they were seeking to leave in favor of more lucrative jobs in the cities. Peasants joined together to raise money and set aside marginally productive land that they then declared to be an industrial park, and advertised for tenants. Using a locally colorful phrase, they sought to 'build a nest on the land left by the ancestors to attract phoenixes' in order to 'head wealth for descendants' (Winnerway 1999, p.2). In 1990 they increased the size of the initial land by almost double, and opened a second, smaller park area. Together they were able to attract 11 companies with foreign investment, related to historic receiving areas for local area residents.

By the beginning of the two-year economic boom in 1992, Winnerway had parlayed early success by diversifying into real estate, trading companies, and agricultural concerns. Joint ventures were actively sought to strengthen its financial capital base and diversify operations. In cooperation with the Dongguan City municipal government, the company also began construction of a bridge to tie Winnerway Commercial Town and Winnerway Industrial Area together in the southeastern section of Dongguan. Clearly, this was a highly successful TVE. In late 1992 Winnerway Industrial Share-Holding Company, Ltd., formed as an internal stock company. Two years later it was listed on the Shenzhen Securities Exchange. Subsidiaries and offices are located throughout coastal China, Hong Kong, and Los Angeles.

Following restructuring in 1996, Winnerway decided to venture into more high technology areas. A joint venture with a Canadian company in 1998 led to formation of the Dongguan Winnerway Bioengineering Company to produce drugs using monoclonal antibody techniques for skin and tumor problems. It now specializes in providing pills, injections, tablets, and capsules for assorted heart maladies and treating hepatitis. Bioengineering efforts in conjunction with the Chinese Academy of Agricultural Sciences also deal with agricultural needs, using microbial research to supply products for 'preventing diseases, promoting growth and increasing production' as well as 'breeding industries' (www.winnerway.com 2002).

Winnerway's operations now extend to companies within the Winnerway Industrial Zone and Commercial Town, stretching over 180 hectares, 40 plants, and 30,000 people. They are best known throughout China as the sponsors of a national championship basketball team, football team, and Dragon Boat squad. Constituent enterprises include CNAIC Winnerway Automobile Company, Ltd., which manufactures completely made-in-China trucks, jeeps, vans, and cars. The pharmaceutical joint venture with a Canadian company holds hope for diversifying from routine assembly work, employing imported high tech machinery. A Japanese corporation headquartered in Hong Kong owns another affiliated company making magnetic recording heads. Winnerway also extends its name to an orchard, restaurants, hotels, office buildings, electronic capacitors, plastics, garments, tape, paint, paper, and sports teams (GWHC 1999).

While the local township residents share stock in their economic hero, management devolves to a small group of consultants. They oversee the operations of 22 fully funded enterprises, 18 holding companies (including pharmaceutical [2], medical technology development [2], bioengineering [3], electronics, electronics [3], cell and gene analysis), 13 joint ventures, and nine companies in an OEM (original equipment manufacturer, selling under another name as supplier) division. The local origins and focus of this hinterland buffer zone provides a striking contrast to the Chang Jiang Delta's traditional and continuing orientation to the world outside China.

Geographically Diffuse Networks

A new tech park in Dongguan opened in 2001 around a PKU funded company, targeted for that locale due to access to information from proximate international companies also involved in computer parts and assembly operations. The intent behind establishment of the 'Dongguan Sci-Tech Park for Famous Universities of China' lay in promoting economic development through innovation spillovers, in a region far better known for its agricultural and low cost assembly operations. Even this advantage appears to be slipping, as Taiwanese PC-related operations move into the Yangtze delta rather than continuing to settle in the Pearl region. A vast, low cost labor pool is not a strong enough attraction for high technology companies; the soil must be suitable for the right kind of grass to grow, that high tech cows need to consume and produce the right milk, as Hershey, Pennsylvania can attest.

Dongguan's Winnerway conglomerate demonstrates several important mechanisms and strategies along its timeline of development from a humble township and village enterprise run by peasants to a model organization with ties to U.S., Canadian, and Hong Kong companies as well as Chinese research institutes on its consolidated industrial and technology parks. It began with pooled funds seeking local businesses. It grew in the next phase thanks to investment via personal network ties from overseas Chinese. Expansion reached out over this network to cooperation with local government, foreign joint ventures, and local Chinese research institutions for the more technology intensive companies. Manufacturing companies draw on the area's large labor pool. Some ventures are more successful than others – motor vehicles, for example, suffer for lack of foreign affiliation and find a smaller market than joint ventures in northern and central China. TVEs such as Winnerway demonstrate that company clusters and technology park schemes can flourish in a township setting, with 'low' as well as 'high' technology. The distinction is between those involved in simple assembly type activities with research-intensive products (PC assembly) and those in a true learning district translating innovative lab bench ideas to new or adapted products. Dongguan clearly characterizes the first type, and fits in a category combining satellite platforms of low tech multinational with locally generated higher technology growth.

Developments in Dongguan also serve to illustrate the predicted evolution of a '*desakota*' zone between an urban area and the rural agricultural hinterland (McGee, 1991; Sit and Yang, 1998; Lin, 2001). Dongguan's spread covers a large area shaped by the influence of Pearl River Delta growth engines such as Hong Kong, Guangzhou, and Shenzhen, as well as 'growth from below' by TVEs, of which Winnerway is the most notable example. This 'metropolitan interlocking region' (Zhou, 1991), to use a term coined by Chinese geographers rather than the '*desakota'* from an Indonesian word for the same Asian growth pattern, is not the sprawl driven by transportation infrastructure extension and amenity housing opportunities in the West. Chinese government urban policy has sought to upgrade small villages to larger towns, and capture the population in place with locally grown economic outlets, as in Dongguan. Whether or not this model can be adopted throughout rural peripheries of Chinese cities is far from certain; clearly Dongguan's regional location provided opportunities not available elsewhere, as well as local leadership and an historical tradition of repatriated capital flowing into the area from more successful overseas descendants. As an example of *desakota* development in the semi-periphery, however, Dongguan shows that it can be done 'with Chinese characteristics'.

References

Department of Geography, Peking University (2000), 'The Field Trip'. Unpublished background Briefing book for the IGU Commission on the Organization of Industrial Space.

Fan, C. (1996), 'Economic Opportunities and Internal Migration: A Case Study of Guangdong Province, China', *Professional Geographer,* vol. 48, pp. 421-449.

Guangdong Winnerway Holdings Corporation (1999), Corporate brochure.

Lin, G. (1999), 'State Policy and Spatial Restructuring in post-Reform China, 1978-1995', *International Journal of Urban and Regional Research*, vol. 23, pp. 670-696.

Lin, G. (2001a), 'Evolving Spatial Form of Urban-Rural Interaction in the Pearl River Delta, China', *Professional Geographer*, vol. 53, pp. 56-70.

Lin, G. (2001b), 'Metropolitan Development in a Transitional Socialist Economy: Spatial Restructuring in the Pearl River Delta, China', *Urban Studies*, vol. 38, pp. 383-406.

Ma, L. and Fan, M. (1994), 'Urbanization From Below: The Growth of Towns in Jiangsu, China', *Urban Studies*, vol. 31, pp. 1625-45.

McGee, T. (1991), 'The Emergence of *desakota* regions in Asia: Expanding a Hypothesis', in N. Ginsburg, B. Koppel and T. McGee (eds.), *The Extended Metropolis: Settlement Transition in Asia,* University of Hawaii Press, Honolulu, pp. 93-108.

National Bureau of Statistics, (ed.) (1999), *Statistical Yearbook of Dongguan*, China Statistics Press.

Oi, J. C. (1999), *Rural China Takes Off: Institutional Foundations of Economic Reform,* University of California Press, Berkeley, CA.

Park, S.O. and Markusen, A. (1995), 'Generalizing New Industrial Districts: A Theoretical Agenda and an Application From a Non-Western Economy' *Environment and Planning A*, vol. 27, pp. 81-104.

Sit, V. F. S. (1998), 'Hong Kong's 'Transferred' Industrialization and Industrial Geography' *Asian Survey*, vol. 38, pp. 880-904.

Sit, V.F.S. and Yang, C. (1998), 'Foreign-Investment-Induced Exo-Urbanization in the Pearl River Delta, China' *Urban Studies*, vol. 34, pp. 647-77.

Smith, C. (10/8/97), 'Municipal-Run Firms Helped Build China; Now, They're Faltering', *Wall Street Journal.*

Statistical Yearbook of Dongguan (2000), *Statistical Yearbook of Dongguan.*

Tong, X. and Wang, J. (2000), 'Global-Local Networking of PC Manufacturing in Dongguan, China', Paper prepared for IGU Commission on the Organization of Industrial Space, Dongguan, China 2000.

Webster, D. (2002), 'On the Edge: Shaping the Future of Peri-urban East Asia', *The Urban Dynamics of East Asia*, Stanford University Institute for International Studies, Asia/Pacific Research Center Discussion Papers, Palo Alto, California.

Williams, H. (2001), 'Property Rights and Legal Reform in Township and Village Enterprises in China', *Asian-Pacific Law & Policy Journal*, vol. 2, pp.227-58.

www.winnerway.com (7/24/02) Accessed.

Zhou, Y. (1991), 'The Metropolitan interlocking Region in China: A Preliminary Hypothesis', in N. Ginsburg, B. Koppel and T. McGee (eds.), *The Extended Metropolis: Settlement Transition in Asia,* University of Hawaii Press.

Chapter 6

Multinational Development Zone: Suzhou

Historic Scenic Setting, Modern Silicon Magnet

Due to its location on the Grand Canal and in proximity to nationally famous Lake Tai, Suzhou rose to prominence earlier than Shanghai, eighty kilometers to its east. A historically wealthy rice growing center within the Yangtze delta, Suzhou's history reaches back to 514 B.C. (Xu, 2000). A major railroad and express highway now connect the two cities. This link is navigable in less than two hours from the China-Singapore Suzhou Industrial Park (CSSIP) on the old city's east side to Shanghai's Hongqiao Airport on the metropolis' western side (Figure 7.1). Through Shanghai, goods shipped to and from Suzhou have easy access to major shipping and processing centers in Shanghai's ports (though some businesses complain that Shanghai loyalists tend to process their hometown's goods quicker than those from neighboring cities).

Magnificent gardens created by scholar-officials who lived in and retired to the scenic city on the banks of the Grand Canal continue to attract both international and domestic tourists and attest to the continuing cultural importance of this 'Venice of the East'. The region also contains a variety of universities and research institutes, including several with technical specialties. Three major industrial districts are located on the outskirts of Suzhou. The Suzhou New District (SND) contains the largest number of companies, around 400 compared to about 190 in the CSSIP. SND lies to the west of Suzhou. The amount of tax paid by companies in CSSIP is much higher, however, reflecting their relatively larger size and prosperity. Suzhou New District sports a mixture of both Chinese and multinational companies, including many relocated out of the city of Suzhou.

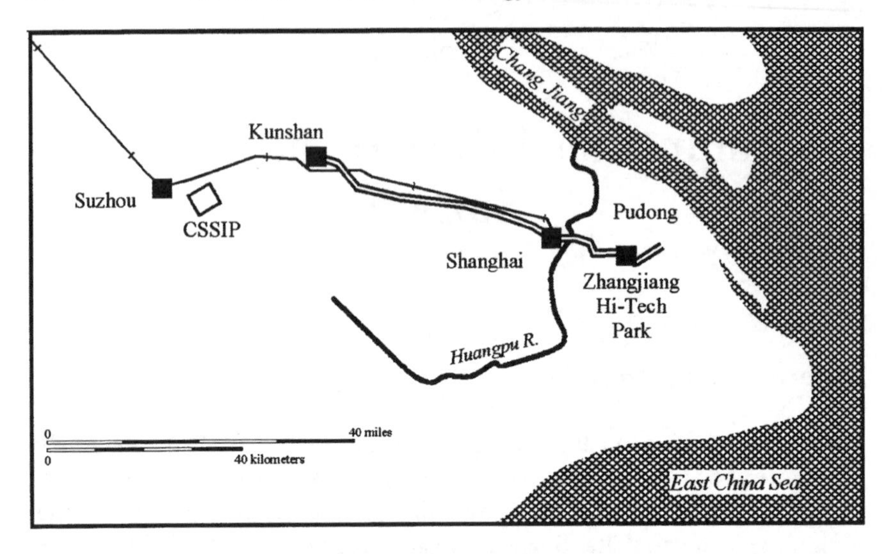

Figure 6.1 Map of Suzhou relative to Shanghai

Kunshan, to the east of CSSIP and closer to Shanghai by 30 kilometers, began with a Taiwanese company. Trains connect the two cities hourly, and the ride itself takes slightly less than an hour by train or non-stop bus (Xin, 2000). Planners anticipate the construction of a light rail link between Suzhou's industrial parks, the city itself, and out to Shanghai. Pudong's new international airport lies almost two hours to the east, and is still not completely linked to the older airport on Shanghai's western edge. Kunshan now contains over 1,000 Chinese companies, such as those from overseas Chinese and Taiwanese. Until recently, these tended to be smaller and less profitable than those in the two other parks. With practical if not formal restrictions on doing business between Chinese and Taiwanese companies lifted, Kunshan attracts suppliers and other production chain elements from that island to serve clients who previously relocated to the mainland. Part of the attraction lies in China's relative willingness to tolerate less environmentally friendly 'dirty' businesses that are also labor intensive in the low-skill range – a sector increasingly abundant with migration from rural areas into more urban settings on the mainland. These jobs were formerly filled by aboriginal people who were willing to work in less desirable sectors of Taiwan's economy, but now government requirements and higher living standards have pushed such activity overseas in Asia's on-going re-enactment of the 'New International Division of Labor' internal to the region. Politics and legal restrictions in Taiwan aside, businesses follow economic imperatives in their migration across the Taiwan Straits. Funneling capital through convenient intermediaries such as Hong Kong, and more recently offshore havens such as the Cayman Islands and British Virgin Islands, Taiwanese investors reportedly sent more than US$2 billion to the mainland from 1998-2000 (Smith, 2001). Sources represent an astounding 90 per cent of Taiwan's high technology companies, as well as those from more labor-intensive sectors.

Not only has much investment come from Taiwan, but also businesses are increasingly migrating from early Taiwanese investment sites in Guangdong province to the Suzhou-Shanghai Chang Jiang corridor. Reasons reportedly include a lingering suspicion of and hostility toward non-Cantonese speakers from outside the province, whereas the central China inhabitants of metropolitan Suzhou speak the same Mandarin dialect *(putong hua)* taught in schools in Taiwan. Inducements to re-locate are also offered by the Jiangsu province and local municipal authorities to make the region competitive with the long-time rival Pearl River delta area. Residents of Taiwan in the Yangtze delta are estimated to range between 250,000 to 300,000. Computer majors produced by China's universities exceed 145,000 annually, at starting salaries three-fourth less than in Taiwan. Unskilled labor to support the top talent level is even more affordable at one-sixth the Taiwanese wage rate (Smith, 2001).

Another trend with long-term implications lies in the return of Yangtze basin natives from more advanced countries such as the United States, forsaking their higher pay and comfortable surroundings for a chance to work in the increasingly available jobs fitting their more advanced skills, in their home region. Shanghai's urban amenities come with a high price tag, so the less expensive but still convenient environs of Suzhou provide an alternative. New residential construction features styles harkening back to classic Chinese patterns, echoing Suzhou's image and appealing to Chinese migrants such as the Taiwanese, with money to invest in these relatively upscale accommodations. These migrants tend to maintain old ties and create new networks in the Yangtze delta, built on language, custom, kinship, and other connective ties to weather uncertainties in the new setting where political, legal, and other local customs and institutions still mark them as outside the mainstream.

Following classic cluster-building theory, suppliers are migrating from Taiwan to serve their old connections in the new Yangtze delta location, as well as providing new orders to established mainland business service companies. These small and medium size companies specialize in more customized production, serving as OEM suppliers for an increasing number of similar sector buyers. The critical mass of accumulating suppliers includes cardboard packing box suppliers, electronics, printed circuit boards, plastic cases, wire, cable, monitors, motherboards, and other facilities from foundries to silicon chips. Such integrated clusters are expensive to construct, and thus represent investor's long-term plans. The most intellectual property-proprietary design sector is the last to relocate to the mainland, while manufacturing and assembly side operations transfer more readily. Research and development centers have been set up within companies, to retain a bit of the proprietary protection of internal innovativeness. Other designers draw on offshore talent and/or local Chinese university hires.

Singapore's Model

Early in their search to find a significant economic factor to keep Singapore afloat in a hostile sea of surrounding countries, the small but ambitious city-state's insightful leaders realized that the island's economic strength lay less in its strategic geographic location as an entrepôt on the Strait of Johore, than in the long-term development potential of a well-educated populace producing for and servicing a high-end global market. As epitomized in its founder, Lee Kuan-yu, Singapore trains excellent engineers – but lacks the sheer number of population to supply the demand its success created (Reddy, 2000). A complementary strategy consisted of attracting foreign direct investment to supply jobs and outside capital. The stable, benevolent-authoritarian government turning out a hard-working, well-educated, English and Chinese-speaking workforce and a modern, dependable and business-friendly infrastructure proved to be the chief lures for scores of foreign multinationals. The island city-state's undisputed success attracted China's attention. Singapore likewise saw a ready supply of well-educated and inexpensive labor in capital-hungry China, looking for friendly tutelage in the ways of market-socialism.

The CSSIP was designed as a business model to provide intellectual as well as physical capital, providing 'customized solutions' from design, project management, fitting out and financing enterprises. Indeed, the design aspires to create a 'work-live-play' environment complete with recreational facilities, business network sessions, seminars, arts and entertainment, convenient stores, schools, and apartments among its billed attractions. These features seem most effective for the Singaporean clientele, however. Non-Singaporean expatriates note the small size of the resident foreign community and the distance to Shanghai, where some send their children to the non-Singaporean foreign school and where the families of others live. Two hours is too far for a daily commute, and a critical community mass is still under construction.

More than 95 per cent of CSSIP's companies are multinational corporations. Most technology transfer takes place within individual companies. Many of the Park's current problems flow from a lack of understanding of geography on the Singapore side: the need for adjusting to a new environment in much larger, less easily controlled China, trusting in the national level government to carry out development agreements while cutting out the province level administrators. The SIP framework reflects policies pursued on the island nation in southeast Asia: support for large MNCs rather than indigenous small or medium companies, stagnation of effort due to a regional economic crunch impinging on prospects in the home country, a branch plant economy (Yeung 2000). Singapore tried streamlining their Park with a one-stop processing set-up, like in Hong Kong influenced SHIP, but Shanghai's bureaucratic influence proved too difficult to overcome. Instead, the one-room setting features a number of desks for each set of forms to be filled out for the panoply of agencies represented.

Differences between the Chinese management of CSSIP and Singapore's representatives shifted in favor of Chinese proposals since the agreed upon flip-flop of ownership in early 2001 from 60 per cent Singapore - 40 per cent Chinese to the opposite distribution. One critical example concerned the relative price of land. Due to its heavy initial investment in extensive, quality infrastructure such that the Park can function independently from its host city, Singapore demanded a fairly high land price, thus improving the competitive advantage of its rival the Singapore New District. From 2001-2002, industrial land price decreased from US$60 per square mile to US$15 per square mile. In the same time period, foreign investment from the United States, Europe and Taiwan increased from US$1billion per year to US$4.7 billion per year. Major areas of investment focused on semiconductor and LCD companies.

Consequently, the rapid sale of almost all available residential area housing pushed the price of residential land up from US$70 per square meter to US$350 per square meter (CSSIP interview).

The SIP management representative, similar to the opinion held by the SHIP representative, insisted that there was no real meaning in the use of 'Industrial' rather than 'Hi-tech Park' in the name, since both predominantly contained companies engaged in high tech sector manufacturing operations. Many companies first set up manufacturing facilities, then added R&D centers to modify product lines. Founded in 1994, SIP's first three years saw basically American and European companies settle. Taiwanese and Overseas Chinese operations increased in the last three years. The type of industries represented includes electronics, pharmaceutical and medical devices, silk, textile, and paint manufacturers (www.cs-sip.com 2001).

When foreign companies assess China as a potential locale for business, particularly as a possible site for sinking considerable capital in manufacturing and research and development facilities, they confront a complex country roughly the same size as the United States, but posing daunting language, customs, legal, and political-economic complexities. Interview subjects repeatedly stated that their location decision was largely based on the location of a joint venture partner and/or the advise of others more experienced, such as a paid consultant. An extremely important competitive feature of CSSIP that foreign companies find very attractive lies in the Development Corporation's role as a matchmaker seeking private partners and suppliers for Park companies. To enhance localization of the supply chain, CSSIP's development office locates suppliers on the outskirts of the Park, keeping the interior for clean, larger corporations.

Singapore-sponsored industrial parks are also in Wuxi (mainly pharmaceutical company focus), Xi'an and Chengdu, but this is the only entity with government-to-government status and purportedly able to approve large projects on its own authority. Difficulties in managing the 'triangular nexus' of firm-host state-origin state relations (Stopford and Strange, 1991) surfaced with Burroughs-Wellcome demand that the government grant it a monopoly on domestic trade in this area as a precondition for completing its 1 billion US$ investment.

The Chinese government sees itself primarily as a logistic and marketing support for companies, drawing the line with BW as the only foreign company permitted in China in this product line. Anticipating the nationalization of SIP, administrative employees are currently sent to

Singapore for training, and then work as assistants to CEOs to prepare for park transition. The Mayor of Suzhou now heads CSSIP, promising more local support following the reversion to Chinese control in 2001. Regional cooperation between Yangtze River delta rivals suffers occasional lapses. Seventy per cent of exports are routed through Shanghai, where port officials who prioritize shipments from their native city rather than upriver Suzhou occasionally delay them.

Future plans exist to expand the park from its present size of 70 sq. km. to 253 sq. km. in two more phases include development of a commercial area as an in-Park downtown, a university area, and more housing. A 'Neighborhood Center' functions as the Park social center. The educational curriculum follows that of Singapore, leading some expatriates to send their children to international schools in Shanghai, such as the one in Qingpu County halfway between the two cities. Plans for developing the CSSIP are ambitious. The goal is to create a city-within-a city (Figure 7.2). Tall retail and office complexes will service nearby companies and their personnel in a modern ('Singapore-esque'?) setting, in contrast to the more aged facilities in downtown Suzhou. The lake will provide an esthetic backdrop, continuing the water-based theme of Suzhou's border on the historic Grand Canal. For the expatriate community in the overwhelmingly foreign corporations within the park, facilities are designed to continue the sustaining bubble notion of all-inclusiveness within the same concentrated area.

Life Science Cluster Firms

The novel notion of locating similar industry companies in close proximity attracted the attention of this research focused on companies in the life science industry. Four of the five companies in this SIP sector were subsequently interviewed. The need to find available plots for companies as they apply to join has diluted this original intent. Significantly for any tests of proximity theory, cooperation between them was greatly influenced by personal preferences and personalities.

Figure 6.2 CSSIP Setting

Becton-Dickinson

The focus of this examination of a typical SIP company falls on Becton Dickinson, a 103-year-old life science giant headquartered in Franklin Lakes, New Jersey. As of 1999 global employees topped 20,000, and capital exceeded US$3.4 billion (Becton Dickinson, 2000). Suzhou employees number over 180, with eight management level positions. The three key worldwide foci are in medication pre-analytical devices, and bioscience. Hypodermic infusion products are a major line. Asia Pacific headquarters are based in Singapore – hence the primary reason for location in SIP. Singapore's 1989 opening represented Becton Dickinson's first plant in Asia; the Suzhou site opened in September 1996.

Due to ISO standard requirements for consistently high quality, products have to be sent to headquarters in the U.S. for testing. All raw materials are imported, with only cartons and packaging produced locally. Managers stated that potential local suppliers required a great deal of training in order for foreign countries from more advanced economies to be willing to do more subcontracting business in China.

Only one out of five laboratories visited by company teams looking for local contacts was deemed suitable under international 'Good Practices' regulations. Emphasis was also placed on the time and expense required to translate all material initially into Chinese, then to send top locally hired Chinese personnel throughout the world for training at various Becton Dickinson locations.

Other Western Life Science Firms

Headquartered in Indianapolis, Indiana, Eli Lilly and Company came to Shanghai in 1928 as the first American pharmaceutical company in China (Bayne, 1999). As did many other Western companies, Lilly left when the communists came to power in 1949, but it eagerly returned under the more open and capitalist-friendly policies of Mao's successors. The company's contacts with China include the Shanghai native who became head of pharmacological research following his discovery that the traditional plant remedy of *mahuang* contained in the popular diet and muscle-enhancing drug ephedrine. At another point, Lilly's Shanghai representative left for Hong Kong during the civil war and became the chief supplier to Chiang Kai-shek for penicillin, the most successful Lilly salesman in the world.

More recently, Lilly had the unfortunate distinction of epitomizing the perils of selling medicine in China when its highly profitable and popular anti-depressant drug Prozac was copied while under patent protection. Lilly is one of three Western life science companies (including Becton Dickinson) on a cul-de-sac in CSSIP, joined around a street corner by giant Burroughs-Wellcome.

The initial attraction for settling in Suzhou, rather than Shanghai, was the English fluency and Western-type interface offered by the Singapore industrial park managers. Like Becton, Lilly seeks to adapt its product to the Chinese market, and struggles to find local suppliers with sufficient quality control for production. The CSSIP management claimed that one of its advertised value-added services consisted of locating reliable suppliers and offering them facilities on the convenient outskirts of the park where they could then provide pooled business services for a (theoretically sufficient) critical mass of park companies. Company interviewees were reluctant to discuss this aspect, but credited park management for trying.

The sophistication and experience of Singaporean managers and their local Chinese counterparts is evident in the design of this park; the problem lies with challenges integrating this design and the experience of Singapore (in a much different setting of a tightly controlled island) with the context of huge, diverse, and bureaucratic China where network relationships are so crucial and lacking with Singaporeans, compared to the overseas Chinese of Guangdong Province.

Network Dynamics: Personal versus Proximal

Labor base issues include the insight that the most important requirement for MNCs is to hire people who can make the right connections, especially with 'flexible' government regulators. Command of English makes for smooth dealing with company administrators, but one must be able to work within a Chinese system whose regulations remain implicit rather than explicit in important areas. The most difficult workers to hire are skilled mechanical operators and experienced managers. Top personnel are sent throughout the world for training at various Becton Dickinson locations. College grads without experience are available, but all are difficult to retain in light of Shanghai's lure. Ties to Shanghai remain numerous and important. Seventy per cent of exports are routed through Shanghai, which sometimes favors items marked as of Shanghai origin (SIPAC 2000). Skilled labor is recruited from Shanghai's top ten universities.

Networking locally is limited, with Becton Dickinson providing friendly advice to some neighbors and new firms in the Park, including some informal mentoring and guided shop floor tours. Personal rather than proximal ties influence these connections, however. No sectoral leadership or networking group has yet formed, despite some attempts in that direction and inquiries from SIP pharmaceutical companies. The American Chamber of Commerce provides discussion and information oriented programs, and the 'Far Eastern Economic Review' sponsors presentations. A university in Nanjing sends students for an internship program, which Becton Dickinson considers a recruitment tool. In the face of cost pressures and other Chinese government restriction on MNCs, the choice may be to consolidate or withdraw, relying on exporting to China from another Asian production base such as the English-speaking Philippine Islands, rather than enduring the hardships of manufacturing locally without compensatory market advantages.

WTO-related concerns should be set within the context of China's dynamic development driving investment such as being the third largest OEM producer of computer parts under leading brands such as Dell and IBM, and the second largest market for computers in the world. The increasing presence of companies acting as upstream suppliers is an unmistakable sign that critical mass is being reached in the Suzhou area's inventory of computer companies, other high technology and even less desirable 'dirty' traditional manufacturing enterprises. All spell economic growth for Suzhou as a technology technopole, within the orbit of its complementary competitor Shanghai.

The initial decision made by foreign manufacturers is whether to come to China and to due what kind of manufacture, then to settle in the central China-Yangtze delta region. The major advantage for being in the CSSIP, according to various interviewees, was the prestigious address due to association with Singapore. Foreign companies in this setting wanted to be with other foreign companies, feeling that it maintained a global business network link and provided a vital bridge to events on that spatial scale. The increasing presence of companies acting as upstream suppliers is an unmistakable sign that critical mass is being reached in the Suzhou area's inventory of computer companies, other high technology and even less desirable 'dirty' traditional manufacturing enterprises. All spell economic growth for Suzhou as a technology technopole, within the orbit of its complementary competitor Shanghai.

Suzhou thus serves as the best example of a deliberately constructed multinational satellite platform. Like its sponsor and model Singapore, the CSSIP seeks out large corporations with headquarters outside China, providing primarily an employment and tax base. While future plans include a new 'downtown' of amenities, a research institute, and range of housing and recreation choices within a greatly expanded park, this is not a technology transfer site so much as a magnet for TNCs within the orbit of Shanghai.

References

Bayne, L. (1999), 'Lilly in China', Unpublished paper, Indianapolis.

Becton Dickinson (1999), 'Growing by Design', Corporate brochure.

Reddy, P. (2000), *Globalization of Corporate R&D, Implications for Innovation Systems in Host Countries*, Routledge, London.

Smith, C. (May 28, 2001) 'City of Silk Becoming Center of Technology', *The New York Times*, C-1, C-2.

Walcott, S. (1998), 'The Indianapolis 'Fortune 500': Lilly and Regional Renaissance', *Environment and Planning A*, vol. 30, pp. 1723-41.

Walcott, S. (2001), 'Growing Global: Learning Locations in the Life Sciences', *Growth and Change*, vol. 32, pp. 511-32.

www.cs-sip.com, (1999-2002).

Xu, X. (2000), *The Chinese City in Space and Time: Suzhou*, University of Hawaii Press, Honolulu.

Yeung, H. (2000), 'Local Politics and Foreign Ventures in China's Transitional Economy, The Political Economy of Singaporean Investments in China, *Political Geography*, vol. 19, pp. 809-40.

Chapter 7

Multinational Learning Zone: Shanghai

Beginnings of a Behemoth

Metropolitan Shanghai functions as the financial, commercial and population center of China, capable of providing the wide range of skills, goods and services otherwise lacking in this newly modernizing country. Of all the locations studied, Shanghai provides the best example of a multinational learning zone, characterized by a large number of foreign firms engaged in manufacturing activities in proximity to Chinese firms, research and development entities, and similar locations both within and outside designated STIPs. Pudong, the new area constructed in the 1990s, particularly illustrates a planned attempt to integrate Western economic conditions in a Chinese setting (Olds, 2001). Beyond and even more important than the presence of mixed MNC and local firms, networks to promote learning such as the American Chamber of Commerce, and various industry cross-cutting, occupation-based organizations, function more effectively here than elsewhere. As predicted by learning district theory in Chapter 2, such a situation holds great promise for Shanghai's continued vibrancy as the commercial center of the country.

Since 1990, the national government of China sought to reinvigorate Shanghai (and its eastern extension of Pudong on the other side of the Huangpu River) as the 'Dragon's Head' of the Chang Jiang (Yangtze River) delta. Shanghai's assigned mission is to pull the country into the 21st century by attracting foreign companies with more developed processes and products into designated Economic and Technology Development Zones (Tan et al, 1996). This chapter examines Shanghai's use of geographically clustered companies as a high technology development tool since the period of 'Reform and Opening' in the early 1990s launched the current transitioning socialist evolution. After a historical review of Shanghai's economic development, the chapter follows the organization outlined in Chapter 2, examining major firms, STIP districts, networks, and global connections in this multinational metropolis.

Shanghai's position as China's traditional industrial base built on its water-based trading channels and highly trained labor pool. Shanghai actively served as a window for dealing with the West since the Qing Dynasty concluded the Treaty of Nanjing to end the Opium War in 1842. During its colonial era (1845-1945), the city was compartmentalized into the densely settled Chinese Old City bordering on extraterritorial British, American (merged with the British), and French districts. Although resentful of foreign intrusions, native employees of the large business interests whose architectural structures remain on the riverside Bund nevertheless enjoyed lucrative economic security unmatched elsewhere in the China. Working for a Western company is part of the present romanticized 1930s nostalgia. During the city's heyday in the 1930s, foreign businesses utilized a core of native businessmen to provide services as 'compradors', bridging the custom gap between the two political-economic worlds of the West and China. Fifty years later, the central government targeted this city for building an economy based on utilization of foreign capital, along the model of Singapore.

Urban design for cities in the new post-1949 People's Republic of China followed a Soviet model specifying distinct zones for particular land use, transforming consumer cities into production centers. Companies grouped close enough to locate worker housing and service facilities adjacent to jobs mingled in an environmentally unsound manner. Cutting down on the negative impacts of commuting also increased residential area pollution. The priority given to developing heavy industry in the first FiveYear Plan led to establishment of several industrial districts on the city's periphery. Five industrial districts and seven satellite towns circle the central city region, with specific functions assigned to each. Six suburbs about ten kilometers from the central district, and six at a distance of 40 km, formed new satellite towns with populations ranging from 50-250,000 (Murphey, 1988). Caohejing focused on electronics and scientific gauges, while Jiading specialized in scientific research and textiles. Minhang was a heavy engineering and machine manufacturing area, Wusong produced iron and steel, Wujing turned out chemical and building materials, Jinshan focused on petrochemicals, and Caohejing in Puxi specialized in electronics (Fung et al, 1992; Ning and Yan, 1995). In 1958 Jiading County sported a new 'Scientific City', including facilities for atomic research and related institutes. The slogan by 1960 was 'science and technology should serve the development of the economy and national defense'. New residential districts to house workers for each industry completed a belt around the central city.

No major metropolitan development occurred from the onset of the Cultural Revolution in 1965 until the late 1970s, when continued crowding bred the next slogan of 'sticking a pin wherever there is space', further compounding the pollution and unhealthy atmosphere of tightly intermixed land usage (He, 1993). Central city finance, trade, communications and commercial networks atrophied in response to the spread of economic activity into the newly created districts encircling the city. The Shanghai municipal government appropriated some central city land, providing it free to industrial and residential developers. Chaotic new patterns flowed from the combination of rapid population growth and push to industrialize (Yeh and Wu, 1995).

In Shanghai's satellite towns, continued emphasis on needs of heavy industry and production over residential amenities meant that urban residents only reluctantly left. China's 'urban bias' – exemplified in the folk phrase 'better a bed in the city than an apartment in the suburbs' – combined with lax record keeping in the all-important household registry to impede distribution of benefits to workers who relocated to the suburbs (He, 1993). Development in the satellite towns focused on attracting new industry, rather than residents, with no particular attempt to alleviate the heavy central city crowding which set world density records. Since services, communication and transportation between satellite towns and Shanghai are poorly linked, population on the outskirts of the metropolis remains relatively low. Integration is better between satellites and rural industrial districts, allowing outlying peripheral industries to take advantage of the lowest labor costs in metropolitan proximity (Ning and Yan, 1995).

Resurgent Region for National Balance

China's rapid emergence onto the global economic and political scene in the late twentieth century followed the ascension of Deng Xiaoping and a major shift in policy. Supplanting the earlier goal of evenly distributing China's development to include interior provinces, the coastal areas were permitted to develop faster, favored by government taxation policy (Fan, 1995). In the immediate post-Mao period, however, Shanghai was considered a pariah for its association with the Gang of Four conservatives. Planning for absorption of Hong Kong in 1997 and seeking to reassure its anxious residents, the central government favored rapid development of Guangdong-Shenzhen directly across the border so Hong Kong residents could witness China's ability to sustain a modern metropolis with striking investment potential.

Shanghai's rejuvenation as a regional competitor to the Pearl River metropolis occurred step-by-step. In 1972, Shanghai scored a high technology coup with the installation of satellite communication stations by RCA for broadcasting Nixon's historic visit. In 1981 Shanghai contributed substantially to the national economy (almost 13 per cent of the gross industrial output); its tax burden supplied over 16 per cent of the national revenue. The next year the State Council formed a Shanghai Economic Region. In 1983 city leadership passed into the hands of technocrats interested in rapid development. Shanghai was designated an open port in 1984. The first U.S.-China joint venture took place with the inauguration of Hewlett-Packard China in 1985. Late in 1986 the city redefined terms such as 'export' and 'technology' in order to grant special enticements to enterprises in these areas. Plans for re-inventing Shanghai's western rural suburb of Pudong were promoted at the national level from 1987-1988.

Prioritizing economic development transformed China in the 1980s. Central control was eased in order to invigorate the economy as a whole. Following Deng's 'Spring Wind' speech in 1992 after his tour of south China, a tilt away from Guangdong-led growth reflected the appearance of Shanghai-affiliated leadership at high central levels (Tian, 1996). Former Shanghai mayors Jiang Zemin and Zhu Rongji rose to political prominence in Beijing. Shanghai's high taxation levels eased considerably, from 90 per cent to 75 per cent, and now 33 per cent. By 1995, the Shanghai-led Chang Jiang delta equaled the Guangdong-led Pearl delta in economic strength. The Shanghai Chinese Communist Party committee formally named five research and development areas: information technology (consisting of microelectrode, computer information and telecommunications), new materials manufacturing (including automobile, bio-environmental, construction, and multifunctional materials), biological technology, advanced manufacturing, and environmental protection. The first three areas were further prioritized as the top High Tech industries for development. Shanghai's proportion of population in manufacturing (secondary) occupations is first among major Chinese cities (Table 7.1).

Table 7.1 Industry sectors in major metropolitan areas

Region	Employment (million)	% Primary Industry	% Secondary Industry	% Tertiary Industry	% Total of S&T
National	705.86	50.1	23.0	26.1	49.1
Beijing	6.21	11.9	33.7	54.4	88.1
Tianjin	4.21	19.6	40.8	39.6	80.4
Shanghai	6.77	13.8	44.3	41.9	86.2
Guangdong	37.6	41.2	26.2	32.6	58.8

Source: China Statistical Yearbook (2000)

Although not among the first cities designated for high technology development (due primarily to political considerations), metropolitan Shanghai now supports a total of seven national level and six city level Economic Technology Development Zones (ETDZs). In order of their establishment, the four development zones profiled in this chapter are the Shanghai-Minhang Economic and Technological Development Zone (SMETDZ, which opened in 1985) located in a satellite city to the southwest of central Shanghai, Shanghai-Caohejing Hi Tech Park (1988) close to Hongqiao Airport within central Shanghai, and Zhangjiang Hi-Tech Park (1992) in Pudong (Figure 7.1).

They vary greatly by size (Table 7.2), with different economic activity industries reflecting a focus chosen by the central authorities to separate the areas, though somewhat artificially. Zhangjiang, for example, discovered the long development time and heavy capital investment necessary for biopharmaceutical development tilted interest toward encouraging IT and software development with a much shorter return on investment turn-around. Inspection trips to India explored the path taken by this Asian leader servicing developed world needs in these two areas. Other Shanghai area High Tech Parks include the Shanghai Manufacturing University Science and Technology Park (1994), the International Fabric Science and Technology City (1994), Jinqiao Manufacturing and Exporting District (applied in 1994 to be recognized as a High Tech park), and Jiading Science and Technology Park, which also broke ground in 1994.

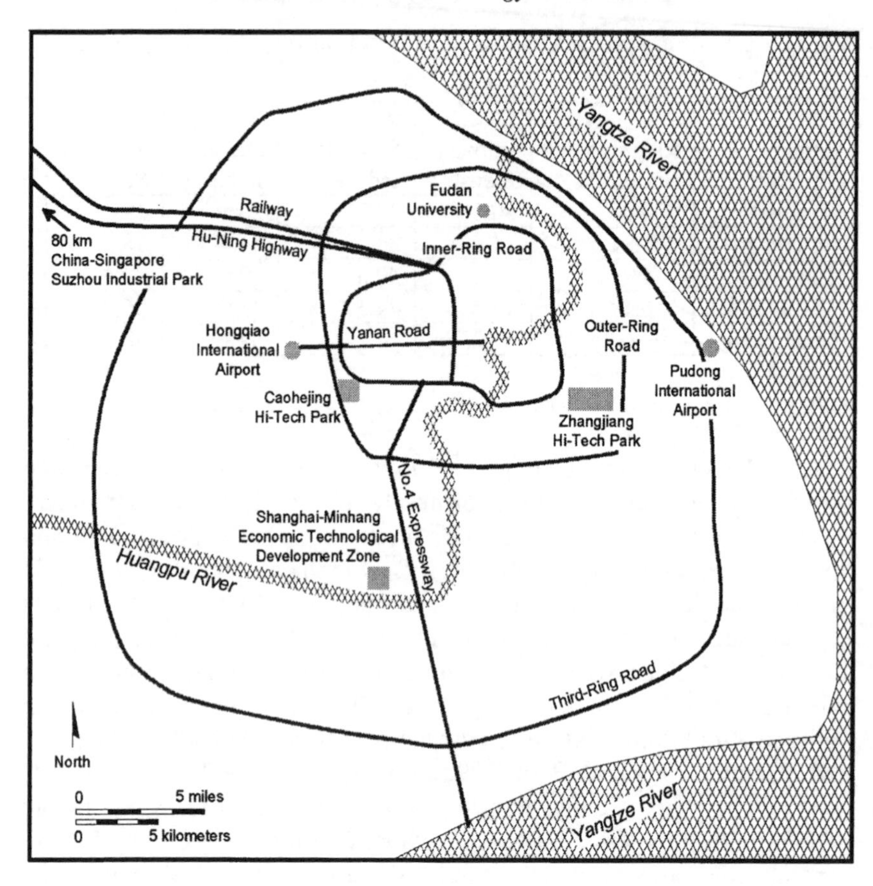

Figure 7.1 Shanghai science and technology industrial parks

Many businesses are directed to their location by the government according to its master plan at that time. The location of foreign High Tech company clusters in metropolitan Shanghai (Figure 7.2) reveals definite concentrations within and outside zones, displaying a combination of the effects of such directed location and the natural tendency for similar companies to cluster together evident in free market countries as well. Administrative offices in an area of high-rise towers comprise the dense concentration along Yenan Road. Caohejing's large number of companies exhibits the effects of its more established High Tech park status, compared to Pudong's larger number of newcomers. This research examines whether Shanghai-Pudong's High Tech districts show sufficient evidence, either in design or execution, of the presence of enough of these factors to indicate the likelihood that China's planners' hopes for these areas will be fulfilled.

Table 7.2 Comparison of metropolitan Shanghai STIPs

Name	Date Open	Km. to CBD	Area (sq. km)	# Companies	% FDI	Focus
Minhang Economic & Technological Development Zone	1985	30	3.5	110	68%	Auto accessories, chemicals, electronic instruments
Caohejing Hi Tech Park	June 1988	11	6 (up to 12)	296	67%	Microelectronics, fiber optics, laser, telecom., bioengineering
Zhangjiang Hi-tech Park	July 1992	16	2.5 (up to 17)	40+	78%	Medicine-based Hi Tech industry
China Singapore Suzhou Industrial Park (outlier)	May 1994	80 (1 ½ hr.)	8 (up to 70)	158	96%	Electrical, chemical & medical engineering, telecom., food

* Center of Shanghai CBD is People's Park

Source: Shanghai Industrial Park Overview, 1998; Shanghai Foreign Economic Relations and Trade Yearbook, 1997; Various Development Corporation publications

Clustering firms in a district draws on advantages that come in general from being in an urban area (improved infrastructure, larger and more diverse labor base, financial instruments, and information access) and advantages from being in close physical proximity to firms with similar production or distribution affinities. Other attributes must be at work for a successful development engine to function. These include leadership that is politically as well as business-connected, continually open information distribution, links to local as well as external suppliers and markets, flexible adaptability to changing conditions, rules and norms that are culturally implicit, recognized and shared, and that foster implementation. The generation of synergy is the most difficult step, especially within China's current framework of information control and

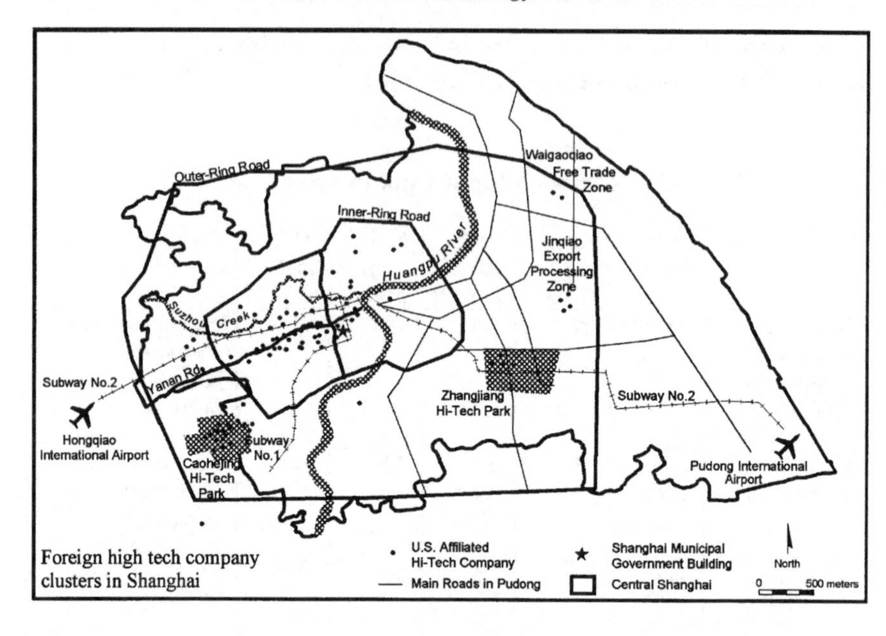

Figure 7.2 Foreign High Tech company clusters in Shanghai

companies from multiple foreign countries. The emerging economic strategy of cooperative competition helps to build critical bridges in this time, with companies looking for a competitive edge breakthrough as a base to build critical mass dominance in some area. The necessity for networks to facilitate information exchange was the object of persistent inquiry in the China interview process. Development districts in Shanghai reflect a unique institutional and cultural context, as well as generally recognizable features.

Shanghai's growth rate and pattern are unique among the world's major rapidly developing municipality's due to the strength of central government control and seesawing policy reflecting the passage of individual leaders with different development policies. As of 1990, central Shanghai consisted of ten urban districts and eight industrial districts comprising 4.4 per cent of the municipality's total land area, into which crowded over seven million people, 55 per cent of the municipality's total population. In the old city alone, population density was 42,900 per sq. km., with an industrial density of 34 per sq. km. - an exceptionally concentrated intermingling in multistory buildings and small local workshops (He, 1993). Adding to its traditional base of financial prominence, Shanghai is currently the center of Chinese international and domestic trade, science and technology, with a third of the nation's scientific research projects (Fung et al., 1992). By the end of 1996, 45 medical joint venture or cooperation companies and ten biotechs were in operation. In the four-year period from 1994-1998, the number of High Tech industries in Shanghai more than doubled (from 353 to 761), as did their total revenue, from US$4.5 billion to US$9.4 billion (Shanghai Statistical Abstract, 1998).

Energy, Brains and Construction Cranes

Currently home to approximately 17 million people, metropolitan Shanghai works around the clock with a skyline of towering construction cranes and flood lit bamboo scaffolding, seeking to remedy problems of over-crowding and outdated infrastructure. Transportation ties with inland China via the Yangtze River and railroad system remains a major strength, with global shipping through ports (Li, 1995). Thousands of industrial sites orbited the central area in three rings, while 70 industrial neighborhoods concentrate in Shanghai's core.

Metropolitan Shanghai sprouted seven state-level, ten municipal-level, and 30 county-level industrial parks since the 1980s as part of the strategy known as Opening and Reform, designed to entice foreign companies with more advanced technology to locate in China as a springboard for the country's economic development. Major high technology sectors in Shanghai, indicated in Table 7.3, are electronic, communication, and medical manufacturing. In general, areas for securing foreign direct investment firms emphasized the auto industry, manufacture of electronic and telecommunications equipment, equipment for power plants, heavy-duty machinery and electronic equipment, petrol and precision chemistry, the iron and steel industry, and manufacture of household appliances (Chesterton, Blumenauer, and Binswanger, 1998).

Table 7.3 Major Shanghai technology industry sectors

Category	Units	Employees (10,000)	Gross value of industrial output (100 million RMB)
Equipment Complex Manufacture	1,498	25.17	558.23
Electronic Information Equipment Manufacture	621	13.88	1,013.09
Biopharmaceutical	421	7.17	196.17
Automobile Manufacture	430	9.56	723.87

Source: Shanghai Statistical Yearbook, 2002

Firms

Interviews with firms in Shanghai's STIPs involved a range of sectors and countries of origin, but almost all companies fit within the Chinese definition of a high technology producer. They also shared some common attributes reflecting the environment in which they operated. Most inputs, both tools and material, were obtained abroad, usually from familiar sources in more developed countries. A fitting phrase would be: 'source globally, assemble locally'. Although tax credits were considered important in the location decision, they were preceded by questions of convenience: to a joint venture partner, to a good port for multimodal goods transit, to residence and other companies – though development of infrastructure for rapid and convenient access blunted this consideration through the late 1990s. The '7 Connections' of infrastructure (water, electricity, sewage, telecommunications, etc) were deemed easy to obtain, as physical rather than human acquisitions. Many companies voiced the intention to expand, citing the development of a critical mass of supporting services that had developed over the past five years. They also voiced awareness of the need to be market responsive rather than production driven. Companies largely settled in the newest place opening up, whether by inducement from developers or government directive. The balance between the push-pull factors depended largely on the business's time of arrival in China in relation to the government's policies on developing a particular area or catering to the desires of a particular company.

Very little technology transfer seemed to be occurring, in part due to concerns about the intellectual property protection (all complained about the lack of enforcement in this area) and in part due to the difficulty of obtaining qualified highly skilled labor in areas such as electrical engineers. It is an 'employer's market' for highly skilled employees, who then need to be trained in the areas of maintaining consistently high quality and proceeding through tasks by sequential steps. Some contract work was being undertaken with top quality technical universities, but more for strategic reasons than R&D.

In Shanghai, the large number of TNCs and the municipal government's pro-business attitude led to a stronger bargaining position for TNCs (Yeung and Li, 2000). As one manager of a large firm put it, companies cluster from a need for a 'big voice in a big country' (interview, 2000). Some sectors were relatively well connected with an industry intranet electronic bulletin board and industry organization (chemical, pharmaceutical). Overall, the opinions echoed those of an earlier study (Nyaw, 1996), finding that Shanghai serves as a premier commercial location for foreign and high technology firms – within a legal and financial national framework that still needs opening up.

German companies form the second largest group of foreign companies in Shanghai, comprising some 620 companies including 320 representative offices, fifteen branch companies, 150 wholly foreign-owned subsidiaries (WFOEs) and 150 joint ventures (JVs) in mid-2001. The major concentration of German firms is in the chemical industry, with a niche specialty in exhibition centers. Germany is known throughout Europe for their large and successful exhibitions organized in Frankfurt and Hanover. The majority of foreign firms located outside of designated development zones indicates gaps in the competitive services offered, as well as the ability of small companies to 'fly below the radar' of government authorities who designate corporate locations. They also benefit from the recent suspension of many requirements for approval of foreign ventures. No longer do they have to be classified as 'high technology' firms, maintain a designated Chinese currency balance sheet (a guard on the export proportion of business relative to domestic market sales), or export a fixed minimum proportion of product.

Less isolated than their country cousins in Taicang, the approximately 2,500 Germans living in Shanghai rely more on local supply chain elements. While a major motivation for entering into a joint venture arrangement with a Chinese company is access to networks of reliable and knowledgeable suppliers, lack of information about how to assess potential partners confounds particularly smaller and more vulnerable foreign firms on this critical point (Strange 1998). Service industry firms catering to German and other clients follow the usual pattern of locating in Puxi.

The price of real estate and the size of facilities available strongly influence location decisions. The Germans plan to make Shanghai their premier Asian exhibition site, further concentrating business attention on this city. One third of German businesses in China are located in metropolitan Shanghai, including the satellite cities of Suzhou (50 miles from Shanghai) and Taicang (30 miles from Shanghai). While industrial parks within Shanghai basically cater to large companies such as Siemens, who brought around 60 affiliated companies with them, Taicang attracts small and medium size companies comprising the bulk of German enterprises and 85 per cent of that country's gross domestic product. The majority of foreign firms located outside of designated development zones indicates gaps in the competitive services offered, as well as the ability of small companies to 'fly below the radar' of government authorities who designate corporate locations. They also benefit from the recent suspension of many requirements for approval of foreign ventures. No longer do they have to be classified as 'high technology' firms, maintain a designated Chinese currency balance sheet (a guard on the export proportion of business relative to domestic market sales), or export a fixed minimum proportion of product.

On the cultural side, services provided by the Delegation of German Industry and Commerce in Shanghai include minicourses for primarily Chinese employees of German firms ranging from language to leadership training, personality type identification, and business etiquette. Typical of Western firms, these courses aim to promote the identification of workers (and particularly expensive, difficult to locate top native management) with the culturally embedded business practices of the firm's headquarter. Part of this familiarization involves a short trip to the headquarter country, though for limited numbers of top employees even in German firms famous for utilizing the apprenticeship method, due in part to the cost but more to the danger of employees opting not to return – or raising their desired salary demands commensurate with their heightened sense of their worth to the company.

Intra-Metropolitan Development Districts

A contrasting picture of the nature of industries in and outside development zones in Shanghai is presented in Table 7.4; note particularly the category of highest proportional representation within zones is for science and technology companies, who also earn more than half of all foreign exchange.

Table 7.4 Location and features of Shanghai High Tech companies

Category	Total	In High Tech Parks*	Other Shanghai Areas
Industries	587	219	368.0
Gross Product (100 million RMB)	526	201	325.0
Science & Tech Input (100 million RMB)	28	13.4	14.6
Employees (10,000)	20.4	4.7	15.7
Foreign Exchange (100 million US$)	10.64	3.9	6.7

*High Tech Parks include six areas so designated by the government

Source: Shanghai Statistical Yearbook of Science and Technology (1998)

The cost of land and facilities at different sites constitutes one factor influencing the location decision of firms seeking to decide between metropolitan Shanghai's various development zones (Table 7.5). A range of prices indicates the calculations of various zone management companies and the power of various companies to seek preferential rates. Prices also vary across time to reflect the relative availability of facilities, in a market that shifted from highly constricted since 1949 to considerably overbuilt in the first flush of development in the early 1990s. Office rents also fluctuated to reflect relative desirability, with Pudong generally commanding higher rates than Puxi.

Measures of the parks, and the new development area of Pudong as a whole, are provided in Tables in the sections that follow. Development in the three zones demonstrates that electronics and telecommunications equipment and engineering sectors performed well in all. Although a tendency exists to inflate output value and deflate profits, number of units and employees generally increased over time. While Zhangjiang had more pharmaceutical companies than the others, this was only half of Pudong's total. Data availability restricts some comparisons. Clearly, the decade of the 1990s provided a dynamic foundation for metropolitan Shanghai's contribution to China's modern development. The following section profiles the evolution and character of Shanghai's three featured High Tech parks.

Table 7.5 Land and building prices in Shanghai zones

Location	Land asking rent (US$/sq.m./mo)	Land asking price (US$/sq.meter)	Building asking rent (US$/sq.m./mo)	Building asking price (US$/sq.meter)
Jinqiao	7.8	$95	$3.75	$300-400
Lujiazui	NA	$550-600	NA	NA
Waig'qiao	8-10	$95-120	$3.75-5	$350-470
Zhangjiang	6	$50-75	$6	$300
Xinzhuang			$2.2-2.9	$96-108

Source: Pudong New Area Administration; Economist IU, 2002

Minhang: Traditional Suburban Production Milieu

Shanghai's first industrial park open to foreign firms was created in 1985 following establishment of Squibb Pharmaceutical's plant in what became the Shanghai Minhang Economic and Technological Development Zone (SMETDZ). Minhang is the earliest developed and smallest zone by size, but has one of the highest value outputs per area (Table 7.6). With a history of opening enterprises in far-flung corners of emerging global markets, Squibb was the first American pharmaceutical firm to set up a joint venture in China when it signed the documents in 1982. The government provided land on favorable terms, promised imminent infrastructure improvements, and created a development zone to entice other foreign ventures in a 1984 competition between Minhang, Caohejing, and Hongqiao (Interview, Squibb 1999).

Although the latter two are much closer to the Shanghai CBD, Minhang sports an established tradition as an industrial city with a good labor skill base. Minhang companies tout the industrial labor base and its location equidistant to Hongqiao Airport and the CBD. In Minhang, ten per cent of the workers commute from Shanghai; many of the managers live in the Hongqiao foreign quarter by the airport. A second, municipal-level industrial park flourishes close to SMETDZ, but the state-supported entity includes four large state-owned power, mechanical, and electrical companies. By 1999, 141 projects in the Zone represented over US$1.75 billion in investments from eighteen countries and regions. Foreign firms such as Coke, Pepsi, Johnson & Johnson, and Xerox maintain facilities in this crowded area (SMUDCL 1999). With land at a good price, and the Chinese requirement that only foreign products made in China can be sold in China, some manufacturers purchased large lots and expanded with additional product components. Minhang's disadvantages lie in its relative distance from Shanghai, despite a broad and largely untravelled highway, and a concentration on consumer goods companies rather than the flashier and more lucrative high technology segment.

Table 7.6 Minhang index of technology development

	1990		1997		2001	
Category	#	Gross output value of industry *	#	Gross output value of industry*	#	Gross output value of industry*
Total	34	108.6	53	10045	93	17,830
Chemistry	9	9.626	9	519	11	2,033
Pharmaceutical	1	10.704	3	1208	3	1,622
Engineering	7	36.088	8	2582	14	3,982
Electrical Equipment & Machinery	1	0.062	2	274	2	395
Electronics & telecom. equip.	2	6.636	1	285	6	3,779
Measuring Devices	1	0.693	2	502	4	908

Source: Shanghai Statistical Yearbook (1991, 1998)

Satellite city resurgence

As in Caohejing, the next development zone established, the industry base is quite varied and Chinese companies substantially outnumber those with foreign connections. The national composition of companies reflects the Chinese policy of attempting to place local enterprises in close geographic proximity to possible sources of foreign technology transfer (Young and Ping, 1997). Companies sometimes maintain a separate administrative and sales office in Shanghai to be closer to higher-class residences and business activity.

Caohejing: Sino-Silicon City Center

Caohejing companies cite infrastructure and transportation convenience, from Jiang Zemin's original (yet unrealized) idea of it as China's Silicon Valley. Caohejing was singled out in 1988 as the first High Tech development center among the coastal Economic Technology Development Zones. Financial, quality of life and landscape amenities were included as enticements in the five sq. km. area only seven km. from the Hongqiao airport and close to an inner ring road. Major industries range from telecommunications, microelectronics, computer software, and synthetic materials to bioengineering (Table 7.7) (Fung et al,. 1992). In March 1997 a 20,000 sq. m. High Tech incubator was completed. Located closer to the center of Shanghai than other zones examined in detail, it is easily proximate to some 20 universities, 100 research institutes, new residential quarters, and city-level water, electricity, sewage, and telecommunications facilities. A Sino-British consortium is building a new technology park by 2010 within the zone.

Caohejing includes almost 300 Chinese companies and development institutes in the park, slightly more than half of the total entities. One third of the remaining businesses have American ties, followed closely by companies from Hong Kong and Japan. Two companies each from Taiwan and England complete the top five, with representation from 11 other nations. It is somewhat difficult to account for all Taiwanese investment, since for political reasons some is routed through Hong Kong. Firms are typically located quite close to each other, behind a landscaped wall whose parking opening is guarded by a security check. This arrangement provides the advantage of conserving expensive city land, reducing distances traveled by employees and managers between buildings, and for employees biking to their apartments located adjacent to the Park.

Caohejing is closest to the Shanghai Central Business District, but is less landscaped, trading convenience for some environmental amenities. Caohejing is a for-profit organization run by a Chinese banking consortium. Businesses are densely packed though landscaped as usual behind perimeter borders, and Park land is expanding. Opportunities for social interaction vary in each location. Food provision is one barometer of social accessibility, providing possible gathering places for interactions. Minhang features separate company cafeterias, while Caohejing offers a variety of eating options in its spacious development center.

Table 7.7 Caohejing index of technology development

	1990		1997		2001	
Category	#	Gross output value of industry*	#	Gross output value of industry*	#	Gross output value of industry*
Total	55	234.521	92	6,562	110	10,714
Chemistry	4	20.392	3	185	8	916
Pharmaceutical	0	0	2	25	3	129
Engineering	6	8.706	14	627	16	676
Electrical equipment & machinery	2	5.937	11	1,141	11	905
Electronics & telecom. equipment	22	156.646	36	3,518	41	5,819
Measuring devices	11	17.550	13	860	15	1,279

- Gross Output Value of Industry and Profit are given in 1,000,000 RMB (8 RMB = $1 US)

Source: Shanghai Statistical Yearbook (1991, 1998)

University Technology Links: Fudan University Example

A critical mass of industry and activities sufficient to promote and sustain networks of High Tech activities is under construction in Shanghai. Recognizing that every investment decision has a spatial component, the goal is for the area's pool of China's highest skilled improved infrastructure and service sector to produce agglomerated economies of scale that will attract foreign direct investment in a 'virtuous cycle' (Qu and Green, 1997). Technically educated Chinese university graduates and professors constitute an under-utilized, under-paid and critical segment of the labor market. Their innovations tend to be utilized far more by local entrepreneurs than by the government-favored foreign sector, thus directing to Chinese companies a major factor for competition based on local innovations (Anselin, 1997). Foreign companies seeking to tap into local talent and mandated to both 'nationalize' their operations while raising the proportion of local research and development conducted would do well to build more bridges with educational powerhouses such as Fudan University, the only prominent Shanghai area universities with technical and applied programs in the sciences and engineering.

Fudan University's Industry Office is a premier example of a university dedicated to 'the nation's economic development' and in turn prioritized for national investment funds. Its numerous partnerships with foreign universities include one with MIT's innovative Sloan Business School. The Genetics Engineering Department is the national Key Lab in this area, thus targeted for special funding and encouragement to collaborate and disseminate results (Pflaker, 1994). The School of Management includes a Sino-Japan International Management and Technology Center. Many corporate and governmental interests outside the university have an interest in ideas produced there. Universities are encouraged to engage in such income-producing ventures in order to supplement waning government funds diverted to the demands of municipal area building construction (Kinoshita, 1995) .

Shanghai area foreign companies draw very little upon Fudan's research abilities, relative to their availability to Chinese companies. Cooperation with Fudan professors was most frequently cited among interview subjects, but only for market assessment purposes as a pooled foreign company undertaking. The University-run Industry Office is involved in 27 fledgling enterprises under its own roof or nearby in Shanghai. Reluctance of new enterprises to leave the university area, thus distancing themselves from access to advice from known professors and familiar labs, is a key location factor (Interview, Director of Office of Technology Transfer, 1999). Western tech transfers exhibit the same pattern of ties to specific companies and the immediate locale, an embeddedness exacerbated by the uncertainties of foreign ventures in China. Computer information technology, biology and medical devices, electric and microelectronics, environmental protection, technology training, and High Tech company services for both Shanghai and Hong Kong companies comprise the top six areas of concentration for university research. The last two are particularly in demand recently, indicative of the gap with the West in all business-related services during the period prior to 1980. The Office sees its role as a hatchery, nurturing fledgling companies for the first two years of their existence as a commercial venture. The University office works with and considered combining with two other nearby technical universities who also nurture starter firms. Outside firms also set up their own research institutes within Fudan, and participate in holding companies with the university. Venture capital remains a problem in a highly under-financially capitalized country rich primarily in human capital. A key consideration for the strategy of Opening and Reform concerned inviting in foreign companies as potential venture capital sources for Chinese start-ups; instead, Hong Kong investors lead the lot in acquiring the most promising small struggling companies (Saywell, 1998). Despite the resources of a business school, marketing and business managers are in short supply, given the new turn to capitalist forms of economic ventures. Professors face pressure to provide innovative technology and personnel, spurring a new teaching emphasis on creative thinking and practical problem-solving applications with a marketable solution.

While the government also furnished funds to Caohejing for its incubator, very little commodifiable activity has yet appeared. The park's 'Innovation Center' supplied some funds for well-known scientists to work in better facilities outside their university, as well as incubator space for students over a three-year period (Interview, Assistant Director of Caohejing High Tech Park, 1999). Fudan University's personnel were particularly aware of the gap between time expectations of funders and the practical difficulties of turning ideas into companies. Companies most frequently hired students who worked with professors engaged in research in their area of product, and hired consultancy services on a project basis. Lack of sufficient students in a particular field such as electrical engineering was noted.

Pudong-Zhangjiang: Global Commercial Center

The longest and historically most important navigable route into the formerly 'sleeping dragon' lies in its mid-section, the Chang Jiang. The

The longest and historically most important navigable route into the formerly 'sleeping dragon' lies in its mid-section, the Chang Jiang. The tip of its delta is the bustling port of Shanghai, and its 'Dragon Head' extension on the western side of the river, the formerly agricultural town of Pudong (Tan, et al., 1996). In a roughly triangular shape, Pudong is delimited by the Chuanyang River on the south and surrounded to the north and east by the Chang Jiang estuary terminating its long journey into the East China Sea. Pudong was eyed as a development pace setter since the early outward-looking self-strengthening plans of Sun Yat-sen in the 1920s (Dai, 1990). Its geographic counterpart across the Huangpu River is Puxi (or 'Pu-west', compared to Pudong, or 'Pu-east'), site of impressive, Western-looking buildings built by occupying European powers on the Bund in the early 1900s.

Inventing a city

Development of Pudong centers on 350 sq. km. of formerly agricultural land, half the size of Singapore and 12 times that of the other fourteen coastal economic development zones (Tan, et al., 1996). The total area of Pudong runs to 522 sq. km, encompassed as its own Special Economic Zone. Major distinctions in Pudong's planned development include a focus on production for foreign export using foreign investment, and incentives targeted to encourage research and development and local linkages, which were largely absent in the past (Khan, 1991). The total area of Pudong is almost 70 per cent of Puxi's size, with 11 per cent of the total city's population. Premier Li Peng declared in April 1990 that Pudong's rapid development was to 'boost the economy of the Chang Jiang Valley', catapulting Shanghai into the forefront of China's financial and trading centers as a vibrant modern urban area (Li, 1995). By 2001, Pudong supplied 22 per cent of Shanghai's GDP, 20 per cent of its exports, and attracted US$30 billion in FDI, with exports in 2001 of US$11.5 billion.

The opening of Pudong for wholly foreign owned financial institutions was a major step drawing money into local businesses (She, et al., 1997). Breakthrough strategies include allowing foreign banks to do local currency exchange transactions. Prevalent modes of doing business are both joint ventures and build-operate-transfer procedures, under which MNCs construct a business that then trains and is ultimately turned over to Chinese nationals (Luo, 1998). Economic regeneration of the Chang Jiang delta region would be largely financed by foreign of Shanghai Municipality for the Encouragement of Foreign Investment seek to establish ground rules for investors lacking in the rest of China, thereby making the city safe for foreign money (Givant, 1991). Incentives for outside funds are crucial, due to the dearth of Chinese investment capital.

Riverside Rivals

A local joke about Pudong holds that perception depends on which bank of the Huangpu River (dividing central Shanghai from Pudong) one stands. While Pudong companies proclaim their clean environment, Puxi calls it empty; what Pudong views as noisy, crowded Puxi on one side, the other declares shows its desirability. Across the busy Huangpu directly facing the historic Bund banks and other colonial-era buildings, the modernistic architecture of Pudong's Lujiazui 'Wall Street' district provides a striking view of New China (Figure 7.3).

Shanghai invested over 310 billion yuan (nearly US$40 billion) in infrastructure improvements during the last decade of the 20th century. Information capacity enhancement includes integrated information pipelines, a broadband IP urban network, a data network, cable television, and a broadband information exchange center. In the year 2000, 15 major projects were completed, including the first phase of Metro Line #2 and the Peal Line light rail, a pedestrian tunnel under the Huangpu River, and natural gas lines to Pudong (www.shanghai.gov.cn, 2002).

Infrastructure construction linking formerly backward, isolated Pudong with bustling Shanghai were the first projects in Pudong's makeover: two bridges spanning the Huangpu and two tunnels underneath. A new international airport and a vital subway connection with Shanghai opened in late 1999. An inland navigation network flows through the Pudong Channel and Chuanyang River. New gas and water works were recently completed, along with additions to a thermal power plant. Unlike many sprawling urban areas in the U.S., the goal for road construction is to link an expanding area to a safety valve engine in the suburbs. Even more important is the design of concentrating the activity of each area. A group of international planners acknowledged the benefits of concentrated economic functions, and sought to develop synergistic interchanges between Pudong's financial, administrative, research, manufacturing and processing centers (Olds, 1997).

Figure 7.3 New Shanghai, Pudong side across from the Bund

Pudong is divided into ten distinct economic and spatial zones:

1. *Lujiazui Financial Area* (5.5 sq. km.) functions as a finance, trade, and commerce center, much like the historic Bund of Shanghai directly across the river. Touted as China's Wall Street, it is China's sole designated finance and trade development zone, and the only place in China where foreign financial institutions are permitted to do business in the local currency (*Renminbi*).
2. *Waigaoqiao Bonded Free Trade Zone* (6.4 sq. km.) serves as China's largest free trade zone, with over 3,500 companies attracted by generous tax forgiveness and other incentives, it incorporates port facilities, warehouses, and a base for natural resource exploration in the East China Sea.
3. *Jinqiao Export Processing Zone* (19 sq. km.) handles high technology, export-oriented 'clean' industries, combining a version of Silicon Valley with Shanghai's traditional strength in textiles (Ge, 1990; Halim, et al., 1996). Of some 302 companies in 1997, including 38 High Techs, 164 represented foreign investments. This zone includes upscale residential areas,

international schools, research institutes and university branches, hospitals, and a separate commercial district for convenient shopping and office buildings. It continues to expand and integrate facilities such as warehouse and greenfield developments, successfully attracting foreigners and their companies. It is located between Lujiazui and Zhangjiang, only nine km from Waigaoqiao.

4. *Sunqiao Modern Agriculture Development Zone* constitutes an important and highly lauded nationally ranked zone for the development of agricultural biotechnology. Expansion of urbanization and rural industrialization into formerly agricultural areas on the periphery of large cities increases the already urgent need for innovative ways to produce more food on less land. Chinese regulations pertaining to cloning and genetically modified organisms (GMOs) are less stringent than in many other countries, in response to the search for a technological solution.
5. *Zhangjiang High Tech Industrial Park* is close to Lujiazui, Shanghai's inner ring road, bus routes to Puxi universities, and the terminus of Pudong's subway link to Puxi. As detailed in a later section, its greenfield remoteness permitted distinctive development that both attracted (due to amenities resembling Western suburbs) and discouraged foreign investors. *Residential Zone, Huaxia Culture and Tourism Zone.*
6. Newer areas include *Wangqiao Industry Zone, Liuli Modern Xinghuo Development Zone. Zhoujiadu Industrial Zone* includes factories for shipbuilding and automotive industry parts. The total development zone area of 37.42 sq. km. absorbs 35 billion yuan of investment (Statistical Yearbook of Shanghai Pudong New Area, 1998).

Powerhouse or Periphery?

Zhangjiang High Tech Park (ZHTP) focuses on pharmaceuticals and electronics, artificial intelligence, lasers, computer software, telecommunications and information transmission. Its three major targets, set by the Shanghai municipal government, are the bio-pharmaceutical industry, the microelectronics and information industry, and optics, mechanics and electronics (Ge, 1990; Tan et al., 1996). Prioritizing medicine reflects both Shanghai's dominance in state-level institutes and universities in this area, and the needs of China's populace (Gross, 1996). Branch campuses from noted engineering powerhouse Jiaotong University, Fudan University (which spawned several companies based on technology developed by its professors and graduates), Shanghai's Medical University, and the Shanghai New Drug Research and Development Center were located in the zone to increase the university-industry proximity.

Zhangjiang High Tech Park realizes its lack of an important on-site university is a weakness, in recognition that knowledge spillovers have a very localized effect. As a compensatory move, bus lines now run frequently between the research universities Fudan and Jiaotong for the ease of employees anchored to those campuses well outside Pudong. Many companies in the Park have their own integrated research labs, or use research and development input from their foreign headquarter location. To spark local company development and scientist-entrepreneur efforts, Zhangjiang seeks to lure Chinese scientists from the San Diego biotech base, hoping they will in turn bring overseas students with their mentors to China, as happened in Taiwan's Hsinchu High Tech Park. Jiading, a satellite city of Shanghai to its northwest, reportedly attracts overseas students to a lower-priced incubator. Consciously following the model of Hsinchu and the U.S.s Research Triangle Park, other national level research facilities have been located or are planned for Zhangjiang, including Fudan University, the Chinese Academy of Sciences Shanghai Branch, and a Shanghai Pudong Software Park that includes both a large building complex and a green area with statues and a lake.

Zhangjiang's overall goal is to span the gap in science and technology between China and advanced countries, developing industries for the 21st century rather than remaining in manufacturing (Fei, 1990; Interviews). Part of the High Tech design is a concentrated 'Medicine Valley' for a group of Chinese, foreign, and joint venture medicine-focused research institutes and pharmaceutical plants. ZHTP's concentration on pharmaceutical manufacturers has led to an increasing realization of and frustration with the long lead time and high cost of the drug discovery, testing, and approval cycle. Of the 79 companies currently listed as occupants, 22 are in the 'Innovation Base' incubator (including two from Fudan). Zhangjiang professes a special responsibility to assist the transformation of these small and medium enterprises, transitioning from government support to market justification. Nineteen companies are in the biotech-pharmaceutical base, ten are in microelectronics, seven are research and development entities, and the balance classified as 'other High Tech'. Zhangjiang sports several centrally located eating facility, which help to combat a sense of isolation in the less inhabited Zone. A webpage of Shanghai metropolitan area events is under construction for Zhangjiang, to assist integration of employees and residents with events.

Government authorities often steer new foreign high technology companies seeking to locate in Shanghai toward considering Zhangjiang. Although the government directed the first company in Zhangjiang there, the large transnational corporation pulled in several other related businesses. Zhangjiang comes closest to an American suburb, and is noticeably less developed than the others with its wide, quiet streets. The presence of both an International and an American school, along with the golf course and luxurious (if speculative) housing area, point to high hopes for future foreign residents, as the country waits to see if its 'Dragon Head' leads development down the High Tech path. Signs outside structures indicate the scale of their sponsoring entities, from the nationally funded Torch Program and Chinese Academy of Science to Shanghai affiliation, Pudong, and Zhangjiang level interests. Old company quarters such as Revlon's are reused as new incubator space for fledgling startups, while Revlon occupies larger quarters within the Park. Entrepreneurs cited the 'open and free attitude toward business' as a critical element supporting new and innovative enterprises. Government support came largely in the form of capital and tax forgiveness, rather than supplying critically needed advice on managing a business and securing future funding. Local university lectures are available, but no leadership appears to be emerging from a particular company, university, or organization.

Networking associations sponsored by the government included the Returned (American) Students Club (*Mei Tong Xue Huei*) and an MBA association, as well as localized industry level groups for software, dot-coms, and the like. Creating a climate conducive to building human relationships (*ren qi*) is particularly important in China, and fostered by company practices such as providing conveniently located and subsidized dormitories for workers, three daily meals in a central cafeteria, and other bonding elements for sharing experiences, supplementing wages. The following Tables portray the relative well being of the major zones in newly developed Pudong. They provide a stronger high technology element than in Shanghai's manufacturing centered economy.

Table 7.8 Zhangjiang High Tech Park, 1997

Category	Unit	Workers	Gross Output Value Industry*	Total Pre-Tax Profits*
Total	35	7576	233.469	4.848
Garment and Other Fiber Product	5	102	0.478	-0.086
Pharmaceutical	4	2359	66.687	2.301
Plastic Product	4	627	10.278	0.759
Nonmetal Mineral Manufacturing	3	1006	11.78	1.981
Electrical Equipment & Machinery	3	348	37.019	-0.816
Electronics & Telecomm Equip **	3	836	51.809	3.975
Measuring Devices	2	94	2.179	0.424

Table 7.9 Lujiazui Finance and Trade Zone, 1997

Category	Unit		Employee (person)		Gross Output Value*		Total Pre-Tax Profits*	
Year	**97**	**93**	**97**	**93**	**97**	**93**	**97**	**93**
Total	212	217	44277	155626	661.67	122 3.6	101.7	94. 2
Textiles	11	19	2902	10884	30.307	65.1	3.8	3.1
Garment and Other Fibers	17	8	1878	1412	21.736	9.2	0.4	0.1
Education & Sports goods	10	19	5179	6962	43.106	39.2	2.9	0.7
Chemical Material	8	8		1729		13.5		1.3
Pharmaceutic al	4	4		5568		43.5		2.2 1
Plastic Product	14	10	2021	2930	17.34	19.6	1.39	- 0.0
Metal Product	18	12	2607	3959	26	54	-4.445	2.9
Machinery Manufacture	20	12	1857	3588	9.604	28.2	-6.38	0.9
Specialized Equip Manu-facturing	11	8	782	3401	6.636	21.2 87	1.074	2.3 97
Transport Equip/Manuf	12	18	10230	23055	270.48	179	65.04	17. 9
Electrical Equip/ Machinery	18	26	5481	5871	109.03	149. 44	25.707	15. 34
Electro/ telecom	11	12	2405	19703	49.346	75.3	8.074	9.8 39

Table 7.10 Jinqiao Export Processing Zone

Category	Unit		Employee (person)		Gross Output Value of Industry*		Total Pre-Tax Profits*	
	97	93	97	93	97	93	97	93
Total	132	72	36715	25848	2117.6	324.2	238.7	23.483
Garment & Fiber Product	11	7	1814	767	35.9	5.9	0.3	0.379
Beverage Manuf.	5	0	1891	0	230.8	0	73.4	0
Chemical Material/ Product	10	7	3107	2914	126.6	33.27	5.7	4.24
Drugs	3	4	1652	1725	48.7	36.7	15.9	0.982
Metal Product	8	4	1866	398	11.00	1.78	6.8	
Original Machinery Manuf.	5	6	1960	3055	17.8	23.62	-0.3	6.541
Specialized Equipment Manuf.	8	6	2635	931	22.5	3.99	-0.8	-0.104
Electrical Equipment/ Machinery	15	6	8403	5850	460.704	141.12		8.254
Electronics telecomm equip **	21	0	5702	0	878.860	0	126.84	0
Measuring Devices	17	1	3512	3001	78.879	4.6	0.16	0.273

Waigaoqiao Bonded Free Trade Zone provides a powerful locational inducement for companies to locate in one of Pudong's development parks, within convenient range of one of China's most liberal import and export facilities. Although the capacity of Shanghai's harbor lacks the deep water draw of Hong Kong's, due to the heavy silt load carried by the Chang Jiang which is dropped along its confluence with the Huangpu River at Wusongkou, a new deep water port will be constructed south of Shanghai to serve ships needing that capacity. Meanwhile, Waigaoqiao provides the efficiency of one-stop processing, duty-free transit, and exemption or postponement of other taxes (Bassolino, 1998). Foreign companies comprise around 95 per cent of the firms in Waigaoqiao. The facility is also noted for ease of importing, necessary for supplying material not otherwise found in China but needed for (particularly high quality, foreign-owned) manufacturers. Its proximity to other high technology-inclined zones established in Pudong in the early 1990s functions as a complementary growth driver, serving as a key logistics node for central China and the Yangtze River delta.

Table 7.11 Waigaoqiao Free Trade Zone

Category	Unit	Employee (person)		Gross Output Value of Industry*		Total Pre-Tax Profits*		
	97	93	97	93	97	93	97	93
Total	82	8	7599	1154	685.87	14.912	18.021	5.918
Textile Industry	8	1	787	58	8.366	1.137	0.903	
Garment Other Fiber Product	11	2	1060	603	10.781	1.658	-0.32	0.096
Chemical Material Product	5	0	226	0	16.518	0	-0.767	0
Plastic Product	5	0	408	0	10.642	0	0.912	0
Metal Product	5		278		7.062		-0.577	
Original Machinery Manuf.	5		348		5.092		-0.79	
Electrical Equip/Mach	7		552		7.998		-0.127	5.106
Electronics & telecomm Equip **	15	1	2556	117	485.684	8.452	17.095	0.137
Measuring Devices	3	1	336	200	106.59	0.736	3.518	

* Gross Output Value of Industry and Profit are given in 1,000,000 RMB (8 RMB = $1 US)

** Outstanding industry in this ETDZ

Chemistry category includes Raw Chemical Material and Chemical Products, Chemical Fiber Manufacturing, Rubber Products, Plastic Products.

Engineering Category includes original machine manufacturing, special purpose equipment manufacturing, transportation equipment manufacturing development zones in Pudong new area include Lujiazui finance and trade zone, Jinqiao export processing zone, Waigaoqiao free trade zone, Zhangjiang High Tech park (not included 1993).

Source: Statistical Yearbook of Shanghai Pudong New Area (1994, 1998)

Human and Fiscal Capital Investments

A focused look at the amount of investment represented by Zhangjiang Hi Tech Park, Shanghai's premier science and technology industrial development area, indicates both the reflection of global financial cycles and the close relationship between foreign and total investment. While the government assists the park management by providing infrastructure basics, foreign firms supply the overwhelming bulk of monetary investment (Figure 7.4).

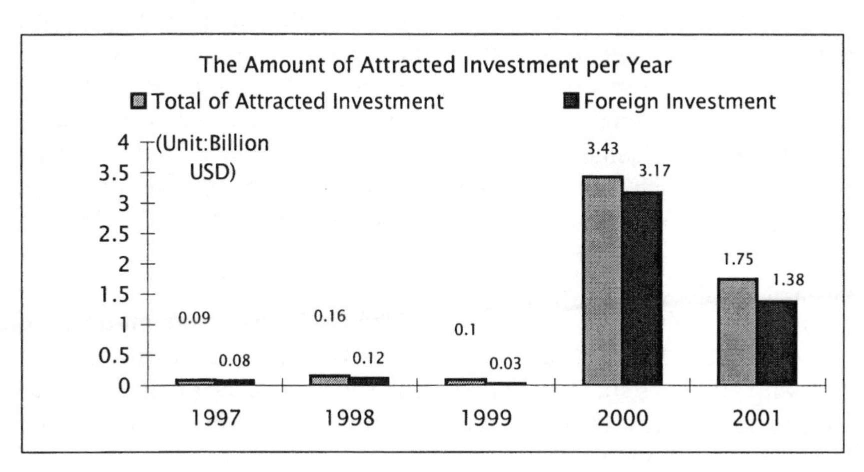

Figure 7.4 Shanghai Zhangjiang Hi Tech Park investment

The nature of the workforce with some degree of higher education reflects the proportion of graduates at different levels in Chinese society: a small elite of PhD's, with a large base at the bachelor's level. Approximately two-thirds of the employees possess no college degree (Table 7.12). Nevertheless, this spatial concentration reflects a large concentration of Shanghai's most highly skilled educated workforce.

Table 7.12 Zhangjiang employees by education

	Total	PhD	MA/MS	BA/BS	Junior College	Others
Worker	32,741	368	1,290	6,303	4,047	20,733

Source: www.zjpark.com

Networks

Although branch plants of some large multinationals conducted business strictly within the firm (such as Lucent-Bell Labs), others reported establishing competitive-cooperative relationships with both traditional rivals and local neighboring plants. Formation of industry-horizontal linkages is crucial for healthy economic functioning in a large country, and an important sign of more relaxed administrative controls in a maturing economy. Of particular note was the observation that companies cited as having a relationship were often directly contiguous, indicating geographically restricted interactions. New relationships are forming with multinational companies from other countries, opening possibilities for future cooperation in other co-located global sites (Hodder, 1990). Shanghai is seen as the preeminent place for new cluster formation. In such a large country with centralized decision-making, companies feel a particular need to band together and speak with one (louder) voice. Some mixing takes place across geographic divides, through organizations such as the 'Foreign-Invested Company General Managers Club' and the countrywide 'China Foreign-Invested Pharmaceutical Association.' Interaction between foreign firms helps to deal with problems such as locating dependable local suppliers. Research cooperation occurs through industry-wide use of Fudan University market studies, for example.

Business Locations

Clustering occurs in a variety of locations throughout metropolitan Shanghai, depending largely upon the type and arrival time of the major company in relation to the opening of the particular development zone. A comparison of the two leading foreign automakers illustrates this point. General Motors is located in Pudong's Jinqiao manufacturing district and shares the same State-owned joint venture partner (the Shanghai Automotive Industry) as Volkswagen, which locates its manufacturing facilities in the suburb of Anting, south of Shanghai. General Motors was permitted to bring many of its global suppliers with it (and indeed forced many of them to come in order to continue supplying the automotive giant). As the earlier arriver, Volkswagen advertises for local suppliers who can meet their time and quality standards and has found them across the metropolitan area. General Motors spent five years trying to pick a suitable location, originally based on the location of a key supplier in

Hangzhou, 100 miles southwest of Shanghai. Two major considerations led to the dissolution of the partnership and General Motors' removal to the Jinqiao district on Pudong. The first was proximity to the duty-free port of Waigaochiao on the northern tip of Pudong, for rapid processing of customs and shipments. The deciding consideration, however, was the desire of its expatriate staff to live in Shanghai given the number and level of amenities available at that time to support a foreign community.

New business possibilities are explored between contiguous companies in several parks, among them Zhangjiang and Jinqiao in Pudong, and Caohejing in Puxi (Walcott and Xiao, 2000). The 'learning district' function of these industrial zones remains confined to tightly defined areas: training employees within the same company, subsidiaries and domestic suppliers of a particular transnational company, Chinese employees of similar Chinese companies who attend district-sponsored presentations, personal friendships of managers who visit each other's shop floors, native universities and their technology spin-offs or graduates.

Residence Locations

Residence locations of expatriates are more clustered than companies in this large city with a population of around seventeen million. Foreigners without fluency in the Chinese language tend to remain in the original large hotel-office-residence-retail settlement of the Shanghai Centre in downtown Shanghai. At 50 stories the tallest structure in Shanghai when it opened in May 1990, this Portman-designed complex includes features such as a theatre, grocery store, nursery school, and health club attractive to long-term residents. Proximity to self-contained shopping (from clothes to travel agencies and Starbucks) and other urban amenities is appealing, as well as the availability of offices an elevator ride away. One of two family centered clusters is in Hongchiao's Gubay district close to the older international airport and on the southwest side of the city. The main American school (or 'bilingual' school) is also conveniently close in the adjacent southern district of Minhang, exerting a pull to the southwest.

A newer center for settlement accompanied the movement of many companies east of the main city to Pudong, following that former agricultural sedimentary deposit delta of the Chiang Jiang's transformation through the 1990s into China's premier showcase for modernization. Hotel and condominium clusters are available in the

Lujiazui financial district and the vicinity of Jinqiao and Zhangjiang Hitech Park zones. Luxurious and extravagantly priced 'villas' have also sprouted in these zones suitable mainly for expatriate salaries.

Residence clusters vary by nationality, age and marital status. European, Japanese, Taiwanese, and other Asian expatriates cluster even more than the American in their residence choices. Bilingual and young foreigners occupy around several Puxi neighborhoods, reflecting both economic and lifestyle choices to be closer to the lively nightlife. Urbane quarters close to Huaihai Road, heart of the former French district, sport Shanghai's second largest glistening high-rise, high priced shopping district (Yeh and Wu, 1996). Due to the nearby location of the Shanghai subway authority headquarters and proximity to Huaihai Road, the main street on one side of a preserved historic 'Model Quarter' contains sports clubs and a lively entertainment area. This particularly area was developed by the local French Catholic church in 1930 at the height of Shanghai's global appeal. Two and three story units are clustered to save space. Leafy courtyards and corridors provide far more than the usual amount of greenery, while wrought iron, slate roofs, and decorative brick impart a uniquely European flavor. Foreign residents fluent in the Chinese language often choose to live here rather than areas of greater non-native concentration such as Hongqiao and Pudong near the two international airports or downtown in or near the Sheraton Centre. An international, or 'bicultural', school is located close by (interviews).

Connections to work and residence were deemed a matter of commute choice based on desired lifestyle, since the completion of the inner ring road, bridges and subways provide manageable accessibility to most parts of the highly concentrated if huge city. Completion of the inner ring road brought the major residence and work clusters of Hongqiao, Pudong, and downtown into comparative proximity. Continuing work on the second ring road along with additional bridges, tunnels and rail connections promises future improvements in structural transportation links.

Social Networks

Functions of networks were explored in interviews with corporate managers, service personnel, and officials with the American, British, and German Chambers of Commerce. The American presence is by far the largest among foreign communities in Shanghai, and the American Chamber (referred to as AmCham) was quickly cited in almost all cases as the premier network organization. In addition to regular membership meetings, AmCham's numerous committees provide a forum for speakers and personal network building. Some committees rigidly restrict membership by nationality, company size, and occupation rank to ensure commonality of focused interests. Services for newcomers include a meeting with five to six established firms who act as experienced mentors to guide the launch into China. Europeans point to the large number of American born and nationalized citizens of Chinese descent, who command unparalleled linguistic, kinship, and physical entrees to local networks as a key asset of American firms. Another major information source was the U.S.-China Business Council, a private organization managed by Americans based in the U.S. but with representative offices in China. They conduct research, publish a highly respected monthly magazine as well as periodic White Papers, and organize other information gatherings. Joint venture partners or industrial park authorities, their functional equivalents, also provide links between the Chinese and foreign business community.

In a city with a small but visible foreign presence, certain places become known for the types of people who patronize them, from a bar frequented by single males of all ages to other entertainment venues for the young managers. As in the home country – only more so due to the 'island effect' of isolation driving social networking for survival – families find network outlets in school, sport, church and charity activities. The boundary between business and social life is fuzzier than in headquarter countries. This situation is a product of the long hours devoted to business, a need to reintegrate with social life, and the need to keep the 'trailing spouse' as happily adjusted and involved as possible, therefore an expatriate organization is highly desirable. Some small business employees emphasize the personal network aspect above that afforded by established channels such as the Chambers, relying on their social instincts to build trust-based relationships and network new opportunities in several respects.

Places for food and drink consumption serve as very popular networking transaction sites. Hotel restaurants initially garnered a good deal of business, since they were familiar to new arrivals (and all who had passed through them at some initial point), menus are in English, the noise level is low, and they are concentrated in easy to reach locations. The hotels' combination of eating and shopping venues recently expanded to encompass several new luxury shopping centers and a string of Starbucks coffee shops along Nanjing Road, referred to by locals as 'Coffee Street'.

Originating in Seattle in 1971, Starbucks opened its first Chinese shop in May 2000 in the upscale shopping district of the former French Concession. Its chief global strategist claims that along with 'the finest coffee' comes 'great people, first-rate music, a comfortable and upbeat meeting place…' (Shanghai Center, 2000, p.6). Indeed, such coffee shops in major business and retail districts provide popular settings for foreigners and often English-speaking Chinese to transact business as well as social affairs. Less time-consuming and less expensive than a full meal event, although Western food such as sandwiches and pizza slices are available along with the sweet pastries more common in American coffee shops, Starbucks imparts an international atmosphere at an elite price, and a notion of proximity to power through cultural consumption.

Connections to headquarters by China-based affiliates were universally bemoaned for the perceived lack of visibility – and appreciated for the discretionary scope provided by a distant office with little interest in or understanding of the personal and business context in which their Asian affiliates operate. The difficulty communicating the differences in living and business contexts is exacerbated by the short time visiting personnel from the headquarter office spent at the foreign location, necessitating a tightly scripted and relatively smooth procession of events from a luxurious hotel base. The challenges of operating in the Chinese environment were seen as almost never officially accommodated. One example was a question about how to get reliable suppliers; where a good Chinese strategy involved enlisting competing suppliers to ensure better service at lower cost, one company rep was restricted by the company policy to relying on only a single supplier. The voluminous set of policies set by headquarters was almost universally felt best ignored, in favor of more novel approaches. A China posting usually falls to a somewhat junior or almost senior staff, anxious to rotate shortly to a more visible final position prior to retirement, thus concerned with establishing a good record for the company in a two to three year window. Shanghai functions as a way station with great risks, and opportunities, maximizing the importance of quickly establishing networks.

Globalization

Until restrictions on foreign investments were eased in early 2000, three criteria guided Chinese acceptance of foreign direct investments (FDI): they should be technologically advanced, possess the potential to generate foreign currency, and combine both urgent need and the inability of China to produce the product domestically (Weidenbaum and Hughes, 1996). Major foreign companies currently represented by facilities in Shanghai include; Lucent, 3M, GE, Ford, Motorola, Medtronic, Revlon and Intel from the U.S.; Germany's Siemens and Boehringer Ingelheim; and Japan's Matsushita, Mitsubishi, Kirin, Shiseido, Toshiba, Sharp, Sony, National, Hitachi and Sankyo. By the year 2000, slightly over half of utilized foreign investment was in manufacturing industries (52 per cent), with the remainder in services (46. = seven per cent), of which a third was invested in real estate (Shanghai Statistical Yearbook, 2001).

Table 7.13 Inflow of foreign direct investment to Shanghai

Year	Amount (in million US$)
1986-91	1,448
1992	1,259
1993	2,318
1994	3,231
1995	3,250
1996	4,716
1997	4,808
1998	3,638
1999	3,048
2000	3,160

Source: Shanghai Statistical Yearbook (2001)

Shanghai's strategies promoting its historically pro-business commercial setting flourished under more lenient central government policies and yielded an accelerating inflow of foreign investment (Table 7.13, Figure 7.5). The numbers shown as coming from Hong Kong, while impressive and in part due to both geographic propinquity and kinship ties, are also widely acknowledged to be exaggerated, masking unacknowledged inflows from Taiwan. For example, though Chinese figures reported Taiwanese investments by the beginning of 2001 at

US$22 billion, the Republic of China announced that it was actually almost five times that figure, representing an entanglement with the mainland economy that disturbed Taiwan's government (Studwell, 2002).

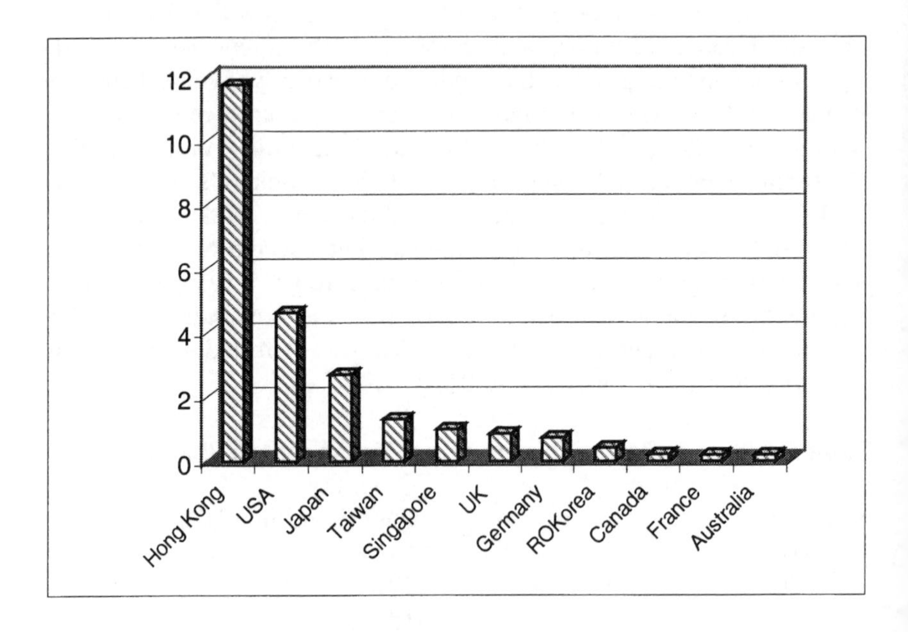

Figure 7.5 Source of foreign direct investment, 1998

Most products reflect the priorities previously set at the national and municipal level for Shanghai to produce the 'six pillar industries' of automobiles (e.g., GM, Volkswagen), telecommunication (Motorola, Bell), oil and chemical facilities, household appliances (Sony, Hitachi), iron and steel, and power station industries, as well as the high technology targets of the biopharmaceutical and computer industry (Ning, 2002). Admittedly, Shanghai usually serves as the assembly site for these companies, who import parts through Waigaochiao's duty free port and the two municipal airports.

Exports are encouraged to earn foreign exchange. Companies varied widely in this respect. Shanghai represents approximately ten per cent of the national market, but markets are highly dispersed. Some companies maintained sales representative offices, working from their homes, throughout the nation. To sell on the Chinese market, key parts can be imported (often through Hong Kong) but the product must be locally manufactured using Chinese labor. Wholly foreign-owned enterprises are required to export half, but joint ventures have no such limitation. Foreign firms vary in their name recognition and local market penetration, but use exports and local market sales to provide balance in times of economic fluctuation. Shanghai thus functions as part of a global commodity chain, occupying a low-cost manufacturing niche as well as a potentially large consumer market (Gereffi, et al., 1994).

A snapshot of Shanghai's foreign links reveals that in 2000 a total of 855 companies showed headquarter locations in the United States, led by the states of California (190) and New York (111); the largest categories were in services such as management consulting (37) and trade promotion (36). Non-service leaders were general industrial equipment (38), electrical components (33), building products (32) and food (31) (U.S. Trade 2000). As a site for patent generation, a frequently used measure of realized innovation capacity, Shanghai scores (an albeit distant) second to Beijing among Chinese cities (Zhou, 2001). As shown in Table 7.14, Shanghai's economy includes a large number of foreign investors, attracts a significant amount of foreign capital, and produces a sizable amount of goods for foreign trade. All increase figures are for the year 2000 compared to 1999. Under the listing of major partners and foreign investment source, offshore entities such as the Cayman Islands largely serve as proxies for Taiwan.

Table 7.14 Shanghai profile

Foreign Enterprises	**Foreign Investment**	**Foreign Trade**
22,270 *FIEs (^41.9%) Total	1,814 Contracts (^23%)	$54.4 bn
2,900 U.S. invested FIE total	$6.4 bn worth of contracts	$14 bn (56%)
32 MN R&D centers	$3.2 bn utilized (^3.7%)	
17 FDI insurers		
Top investors (amount):	**Top partners** (2000):	
Cayman Isl. ($2.5 bn)	Japan ($13.1 bn)	
Hong Kong (940 mn)	U.S. (9.6 bn, ^32%)	
Japan (705 mn)	H.K. (4.0 bn)	
U.S. (508 mn)	Germany (3.6 bn)	
Virgin Isl. (540 mn)	S. Korea (3.1 bn)	
	Europe (7.8 bn, ^48%)	

*Foreign invested entity

Source: U.S. China Business Council, Statistical Yearbook of China, 2000

Clearly, Shanghai participates in the global economy at a very high level within China, attracting considerable attention from foreign companies operating and investing in as well as trading with China.

Metropolitan Shanghai summary

Shanghai amply illustrates the relationship between the attraction of advanced foreign firms and growing local strength. Vernon's (1966) product cycle model illustrating the global location shifts of manufacturing originating in an advanced country has its reverse counterpart in Shanghai as a location capable of furnishing both a low cost production and large domestic market site (Table 7.15). The locational advantages of Shanghai come from proximity to logistics and transportation facilities, funding opportunities and large foreign companies looking for investment openings for acquiring promising local businesses (whether to learn from or eliminate as a rival). Connection to the outside world, so important in recently emerging China, is widely and correctly seen as better in this traditional trading city and 'information port' offering access to the Chang Jiang delta. Competition with the Pearl River delta's Shenzhen-Guangdong-Hong Kong is perceived – and reciprocated. Metropolitan Shanghai trails only Guangdong and Jiangsu provinces (the former due to its Hong Kong proximity, the latter due to its Shanghai proximity) in the amount of foreign investment utilized in the year 2001 (US$3.6 billion).

Table 7.15 Manufacturing location oscillation of global sites

	Advanced Country Cycle	**Developing Country**
Stage 1	Innovate, make, sell at home;	Make foreign product, export;
Stage 2	Lower cost, increase amount: Make abroad, sell at home and abroad;	Make foreign products, sell at home & abroad;
Stage 3	Lower cost & demand: Make abroad, sell at home	Make own products, sell at home & abroad

As a key participant in the global production chain, Shanghai currently serves as a major manufacturing site for foreign products that are then exported for sale, usually to the country where they were designed (from Stage 1 to Stage 2 of the advanced country model, Stage 1 for the developing country). Other Asian country's express concern that China's success at this threatens the future "hollowing out" of their neighbor's low cost manufacturing for export potential. Foreign companies eagerly anticipate the maturation of the Chinese economy to Stage 2, dreaming of China's workers earning enough to advance to Henry Ford's goal of creating middle class consumers fuelling a future surge in production. The goal of China's STIP planners lies in Stage 3, however: local design of innovative products that China can produce and sell to itself, such as the Legend computers and character type setting printers. Stage 3, wherein foreign manufacturers make products abroad for the cost savings in production, but sell basically at home, represents this point in the process.

The capitalist turn provided many learning experiences for both Chinese officials and foreign companies. Positioning itself economically for the next century is the host country's underlying theme. Foreign companies communicated that this requires ordered steps: manufacturing, testing, and then the R&D its hosts eagerly desire. New approaches are introduced via exposure to foreign procedures, such as systemic and creative thinking that encourages innovative problem solving. Companies also report a switch from the earlier production-driven sales 'push' to a more market-sensitive demand 'pull' for production. Japanese companies outfit new factories with globally supplied equipment for turnkey facilities. Companies widely institute full-scale training programs that stress quality control as well as familiarity with the whole production system to understand one's part.

Localization policies promote workers from inside the company to assume top positions formerly held by more foreigners, transitioning to native staff operations. The test of practical and affordable innovative commodities is the marketplace, reflecting both manufacturing and marketing processes as well as the original idea. The comparative advantage of a nation is how research and development is integrated into the rest of the production system. China seeks to address this challenge by upgrading Shanghai's share of critical infrastructure to capture higher-level branch investments. Clustering is clearly accepted as an effective spatial technique, from soliciting High Tech industries in Minhang and Caohejing, to specific industry targets in Zhangjiang.

Overall, Shanghai's strategies promoting its historically pro-business commercial setting flourished under more lenient central government policies and yielded an accelerating inflow of foreign investment

The arrival in the year 2000 of several highly-heralded large integrated circuit companies (headquartered in Taiwan but referred to as 'Cayman Island' companies due to regulations in Taiwan prohibiting their relocation to the mainland) in Zhangjiang points to two several facets of realism impacting planning targets. The central government first invited multinational companies to China in the hope that local companies and employees would benefit from technology transfer from more advanced firms. Their subsequent disappointment with the rate and quality of this localized learning accelerated attempts to encourage the development of Chinese intellectual property in incubators and dedicated park areas. Chinese medicine and software joined foreign pharmaceuticals as targets of Zhangjiang's attraction image. Realization of the length of time it takes to develop such companies and technologies led to a sobering reevaluation, compounded with the temptation to take what opportunities presented themselves in the form of the giant Taiwanese computer chip manufacturers. Chinese planners then decided that foreign manufacturers of High Tech products were valuable occupants of STIPs after all. They also declared the goal (now largely realized) of using these firms as the engine for generating over twenty chip design labs as a higher order complement to the fabrication and assembly functions that preceded them.

Shanghai's Economic Technology Development Zones are local industrial districts set within geographically expanding global corporate and urban networks, with specifically Chinese characteristics. Both previous patterns of globally allied co-located companies are replicated and new relationships forged based on experiences as contiguous corporation in the same park. In both cases, local geography plays a role on the world stage. Traditional districts utilize shared pools of geographically fixed inputs such as labor, infrastructure, tax breaks and tertiary services, low transportation and transaction costs, and a local culture of cooperation. The cooperative-competitive culture is less evident in the developing world model, relying on government-directed resources that put future hopes in previously successful baskets. Areas where entrepreneurship, innovation, and openness to foreign technological adaptation worked in the past form the leading edge for a nation seeking to achieve its global niche.

In the context of a centrally controlled State, the location decision factor is highly constrained. Companies must make the most of wherever they are permitted to locate. This induces multinational interest coalitions based on shared occupation, industry, and location characteristics. Particularly Chinese problems include the re-assertion of traditional localism, resulting in the proliferation of both companies and development parks (established by national, provincial, county and municipal entities) that threaten to inefficiently drain resources while waiting for market corrections to winnow the field. The urge to centralize choices involving allocation of scarce capital by constricting investment targets of venture capitalists is a reassertion of central control that could hamper investment of needed private funds. Price pressure is another point of contention between companies claiming better quality and higher R&D products justifies a bigger profit margin than the government is willing to allow. Too much of a profit squeeze, however, could make China a less desirable location in the global city competition for businesses.

A great deal of variety was noted in transportation infrastructure from one location to another. Assignment of companies to a Zone removed this decision criterion, while promises of infrastructure remedies to come and actual progress in this regard relieved complaints in the long run. The closer to Shanghai the better, was the general consensus. The 'Seven Connections' (electricity, gas, fiber optic network, highway, sewer, telecommunications, water) were all provided for adequately. The prevalence of company buses to bring in workers indicated the overburdened nature of Shanghai's public transportation networks, and the importance of connections to the new subway system. In an attempt to head off acquisitions of motorbikes and private cars, remedies are in the works for roads overcrowded with bicycles and highly unpredictable bus schedules due to congested traffic. Over the next several decades, a combination of subway and elevated light rail lines are planned to result in 21 metropolitan mass transportation connections. Five bridges and seven tunnels are also planned to connect Pudong with Puxi (American Chamber of Commerce, 2001). Asia's first operational magnetic levitation rail line is planned to connect the new international airport in Pudong to the High Tech park in Zhangjiang, a fairly short distance for such speed. Roads in Zhangjiang are usually so quietly suburban that some interviewees expressed a concern as to the availability of otherwise ubiquitous taxis. The section of the Third Ring Road connecting Minhang to the airport was a marvel of open highway space – at least until completion of the other elevated portions. Companies had no problem with globally sourcing inputs.

In general, Chinese workers greatly prefer employment in foreign firms to domestic ones, so the labor pool is competitively highly skilled, abundant and inexpensive. Companies provide buses to pick up workers around the metro area, while every Zone is feverishly constructing on-site housing. This spatial re-concentration represents a successor to the work unit housing provision. Luxury housing adjacent to a golf course, developed by a Taiwanese company, and the new Pudong international airport, seeks to lure expatriate managers from their Hongqiao quarters close to the current airport. Zhangjiang also advertises the existence of Zone-specific international schools.

Lack of linkages with local suppliers, due to low quality control and absorptive capacity of the local economy, inhibits technology transfer. A large gap remains between FDIs with offshore High Tech and large in-house production and small native R&D firms struggling to survive long enough to produce. Several local successes demonstrate the need for innovative products targeted to the local market. Shanghai's utilization of technology parks as 'privileged space' designed to attract High Tech firms for the purpose of jump-starting economic development demonstrates that most active innovating regions are within the world's major metropolitan areas (Castells and Hall, 1994).

Each Zone had its own development organization that was credited by companies with acting as an effective voice on their behalf concerning issues with other governmental authorities. The organization generally owns the land, installs infrastructure, and rents or sells facilities to companies. They also handle arrangements with customs, taxes, banks, license registration, contracts, and environmental issues. All were considered eager to be of service and address corporate needs.

Emerging trends demonstrate evidence of the requisite adaptability. The need to transcend the branch plant stage by cultivating R&D-based industries with future regional potential is clearly acknowledged. The urban spatial form of Shanghai evidences rapid reclustering, particularly in outlying areas, based on new demands for workers in development parks and the construction of sorely needed new housing within those zones. Internationally managed and occupied zones further rural industrialization and integration of Shanghai's satellite cities in the metro urban system. While government constricts location choices, as in the regional analysis framework, the manufacturing zone framework provides a fuller explanation for internal spatial dynamics influenced by zone and industry leaders. Provision of large apartment complexes for zone workers will reshape urban morphology, re-clustering new residences with new job sites.

Companies examined in Shanghai's High Tech parks and industrial zones continue networking processes and competitive-cooperative arrangements found elsewhere. New partnerships are forged within the same industries and in the same Parks, reflecting the closest geographic contiguity. Relationships are created across industries by functions (e.g. CFOs). Major corporations attract global and local supply chain components, located within the same Park or metro area. The urge to agglomerate is particularly fueled by challenges of an unfamiliar, evolving market with vast potential. Information flows must be open enough so that what is produced in the laboratory corresponds to market needs, and/or a market can be created for cutting edge laboratory products. A critical mass of companies and labor pool is generated by a self-sustaining number of both firms and employees, drawn by or created by their predecessors in a particular place. The particular location where an observable High Tech agglomeration occurs is usually based on a real estate configuration.

Several provisions in Shanghai's tenth 'Five Year Plan' demonstrate the perceived importance of continuing efforts to raise the technology level of Chinese economic efforts – the key impetus to formation of STIPs – by 'going all out to upgrade and optimize the industrial structure, emphasizing the role of science and technology progress and informatization'. A two-way development road is envisioned: both 'attracting [companies] from overseas' and 'branching out overseas' to develop as a global economic hub. Urban development will increasingly shift to rural areas such as Pudong and satellite city semi-peripheries, taking account of booming economic zones such as in Taicang. 'Science and technology progress and management renovation' will be speeded up, and exporting added to importing as an emphasis (www.shanghai.gov.cn, 2002). The process of transforming a built environment to an innovative environment also needs to be constructed. This chapter examined to what extent such a process is underway in China's most likely location.

References

American Chamber of Commerce (February 2001), 'Getting Shanghai on the Right Track', *AmChat*, Shanghai, pp. 5-6.

Anselin, L. (1997), 'Local Geographic Spillovers Between University Research and High Tech Innovations', *Journal of Urban Economics*, vol. 42, pp. 422-48.

Bassolino, F. (June 1998), 'Why Waigaoqiao?', *The China Business Review*, vol. 3.

Castells, M. (1992), 'Four Asian Tigers with a Dragon Head' in Applebaum R. and Henderson J. (eds.), *States and Development in the Asian Pacific Rim*, Sage Publications, Newbury Park.

Castells, M. and Hall, P. (1994), *Technopoles of the World: The Making of 21st Century Industrial Complexes*, Routledge, London.

Chesterton, Blumenauer, and Binswanger, Industrial Services Division (1998), 'Shanghai Industrial Park Overview', unpublished document, Shanghai.

Dai, G. (1998), 'Shanghai's Pudong Project in Full Swing', *Beijing Review* vol. 33, pp. 20-4.

Economist Intelligence Unit Limited (2002), *China Hand*, The Economist Publishing Ltd

Fan, C. (1995), 'Of Belts and Ladders: State Policy and Uneven Regional Development in Post-Mao China', *Annals of the Association of American Geographers*, vol. 85, pp.421-49.

Fei, X. (1990), 'Turning Shanghai Into a Mainland Hong Kong', *Beijing Review*, vol.33, pp. 25-7.

Fung, K., Yan, Z. and Ning, Y. (1992), 'Shanghai: China's World City' in Y. Yeung and X. Hu (eds.), *China's Coastal Cities: Catalysts for Modernization*, University of Hawaii Press, Honolulu, pp.124-52.

Ge, W. (1990), 'Rules Add to Pudong's Appeal to Investors', *Beijing Review*, vol. 33, pp. 16-9.

Gereffi, G, Korzeniewicz, M. and Korzeniewicz, R. (1994), 'Introduction: Global Commodity Chains' in G. Gereffi, M. Korzeniewicz, and R. Korzeniewicz (eds.), *Commodity Chains and Global Capitalism*, Greenwood Press, Westport, Connecticut.

Givant, N. (1991), 'Putting Pudong in Perspective', *The China Business Review*, vol. 18 p. 30-2.

Gross, A. (1996), 'Asia's Aging Population Creates Opportunities for U.S. Firms', Medical Device and Diagnostic Industry Magazine.

He, X. (1993), 'Development of Pudong and Optimization of Urban Area Structure in Shanghai, Chinese Environment and Development, vol. 4, pp. 68-88.

Hodder, R. (1990), 'China's Industry-Horizontal Linkages in Shanghai', *Transactions of the Institute of British Geographers*, vol. 15, pp. 487-503.

Khan, Z. (1991), *Patterns of Direct Foreign Investment in China*, The World Bank, Washington, D.C.

Kinoshita, J. (1995), 'Government Focuses Funds and Hopes on Elite Teams', *Science*, vol. 270, pp. 1137-9.

Li, N. (1995), 'Pudong - Full of Hope', *Beijing Review*, vol. 38, pp. 10-4.

Luo, Y. (1998), International Investment Strategies in the People's Republic of China, Ashgate, Aldershot, UK

Murphey, R. (1988), 'Shanghai', in M. Dogan and J. Kasarda (eds.), *Mega-Cities*, Sage Publications, Newbury Park, CA, pps.157-83.

Ning, Y. (2002), 'Globalization and the Sustainable Development of Shanghai', in F. C. Lo, P. J. Marcotullio (eds.), *Globalization and the Sustainability of Cities in the Asia Pacific Region*, United Nations Press, Tokyo.

Ning, Y. and Yan, Z. (1995), 'The Changing Industrial and Spatial Structure in Shanghai', *Urban Geography*, vol. 16, pp. 577-94.

Nyaw, M. (1996), 'Investment Environment Perceptions of Overseas Investors of Foreign-funded Industrial Firms', in Yeung, Y. and Sung, Y. (eds.), *Shanghai*, Chinese University Press, Hong Kong.

Olds, K. (1997), 'Globalizing Shanghai: The 'Global Intelligence Corps' and the Building of Pudong', *Cities*, vol. 14, pp. 109-23.

Olds, K. (2001), *Globalization and Urban Change: Capital, Culture, and Pacific Rim Mega-Projects*, Oxford University Press, Oxford.

Plafker, T. (1994), 'Shanghai Enlists Scientists to Foster Economic Growth', Science, vol. 265, pp. 866-7.

Qu, T. and Green, M. (1997), *Chinese Foreign Direct Investment: A Subnational Perspective on Location.* Ashgate, Aldershot, UK.

Saywell, T. (9/10/98), 'High Stakes: Foreign Venture Capital Spurs China's Hi Tech Start-Ups', *Far Eastern Economic Review*, vol. 161, pp. 66-7.

Shanghai Center (2000), 'Starbucks Has Landed in Shanghai', *Shanghai Centre Tenth Anniversary Issue,* Greenlight Studio, Shanghai.

Shanghai Minhang United Development Co., Ltd. (1999), 'Investment Guide: Shanghai Minhang Economic and Technological Development Zone 1998-1999', Shanghai.

Shanghai Municipal Statistics Bureau (2000), *Statistical Yearbook of Shanghai, 2000*, China Statistics Press.

Shanghai Statistical Abstract of Science and Industry (1998), Shanghai Statistical Publishing House, Shanghai.

She, Z., Xu, G. and Linge, G. (1997), 'The Head and Tail of the Dragon: Shanghai and its Economic Hinterland', in G. Linge (ed.), China's New Spatial Economy: *Heading Towards 2020*, Oxford University Press, New York, pp. 98-122.

State Statistical Bureau, People's Republic of China (1998), *China Statistical Yearbook*, Series 17, No. 17. China Statistical Publishing House, Beijing.

Statistical Yearbook of Shanghai Pudong New Area (1998), Shanghai Statistical Publishing House, Shanghai.

Studwell, J. (2002), *The China Dream: The Quest for the Last Great Untapped Market on Earth,* Atlantic Monthly Press, New York.

Tan, C., Chong, C., and Tan, K. (1996), 'Shanghai's Economic and Trade Zones', in Meng, T.T., Meng, L.A., Williams, J.J., Yong C. and Shi, Y. (eds.), *Business Opportunities in the Chang Jiang River Delta, China*, Nanyang Technological. University Singapore, pp. 73-90.

Tian, G. (1996), *Shanghai's Role in the Economic Development of China: Reform of Foreign Trade and Investment*, Praeger Publishers, Westport, Connecticut.

U.S.-China Business Council (2001), 'Shanghai Snapshots', Unpublished study.

U.S. Trade Council (2000), 'U.S. Companies in Shanghai and East China', Unpublished database.

Vernon, R. (1966), 'International Investment and International Trade in the Product Cycle', *Quarterly Journal of Economics*, vol. 80, pp. 190-207.

Walcott, S. and Xiao, W. (2000), 'High Tech Parks and Development Zones in Metropolitan Shanghai: From the Industrial to the Information Age, *Asia Geography*, vol. 19, pp. 157-79.

Weidenbaum, M. and Hughes, S. (1996), *The Bamboo Network: How Expatriate Chinese Entrepreneurs Are Creating a New Economic Superpower in Asia*, The Free Press, New York.

Yeh, A. and Wu, F. (1995), 'Internal Structure of Chinese Cities in the Midst of Economic Reform', *Urban Geography*, vol. 16, pp. 521-54.

Yeung, Y. and Li, S. (2000), 'Transnational Corporations and Local Embeddedness: Company Case Studies from Shanghai, China', *The Professional Geographer*, vol. 52, pp. 624-35.

Young, S. and Ping, L. (1997), 'Technology Transfer to China Through Foreign Direct Investment', *Regional Studies*, vol. 31, pp. 669-79.

Chapter 8

Local Innovation Learning Zone: Beijing

A Capital Setting

In 1996, at an APEC conference held in Canada, Premier Jiang Zemin set out China's vision for the future with the proclamation that 'the most important creation in this century with respect to technology was constructing science and technology parks' (CSSY, 2000). The model of an industrial district located in a developing world country separate and often far from the countries where many of its largest and most successful companies are headquartered does not fit the reality of Beijing's science and technology parks. Rather than serving primarily as a satellite site for branch operations or (worse, since less economically dynamic) representative offices for major multinational firms, High Tech districts in China's capital city demonstrate the intellectual resources and centralized power of this large country. Many of the most dynamic firms are homegrown, locally staffed and embedded, demonstrating knowledge spillovers from nearby research institutes and the nation's top universities. In an extension of Park and Markusen's (1995) 'satellite industrial district' category, Wang and Wang (1998) refer to China's 'new tech cluster[s]' as local innovation learning zones where new indigenous knowledge and products are being developed. According to Markusen's profit cycle model (1985), these start up companies constitute the first step in a cycle for energizing economic-based regional growth. The difference in China, particularly in Beijing, is that incentives and opportunities crucial to transforming ideas into companies (Hotz-Hart, 2000) are provided only under the auspices of and with the encouragement of the central government.

As the center of political power since the founding of the People's Republic in late 1949, Beijing also served as the focus location for science and technology development through the mid-1960s. By 1965, a total of 55 universities and research institutes were located in Beijing, concentrated principally in the northwestern district known as Zhongguancun (Figure 8.1).

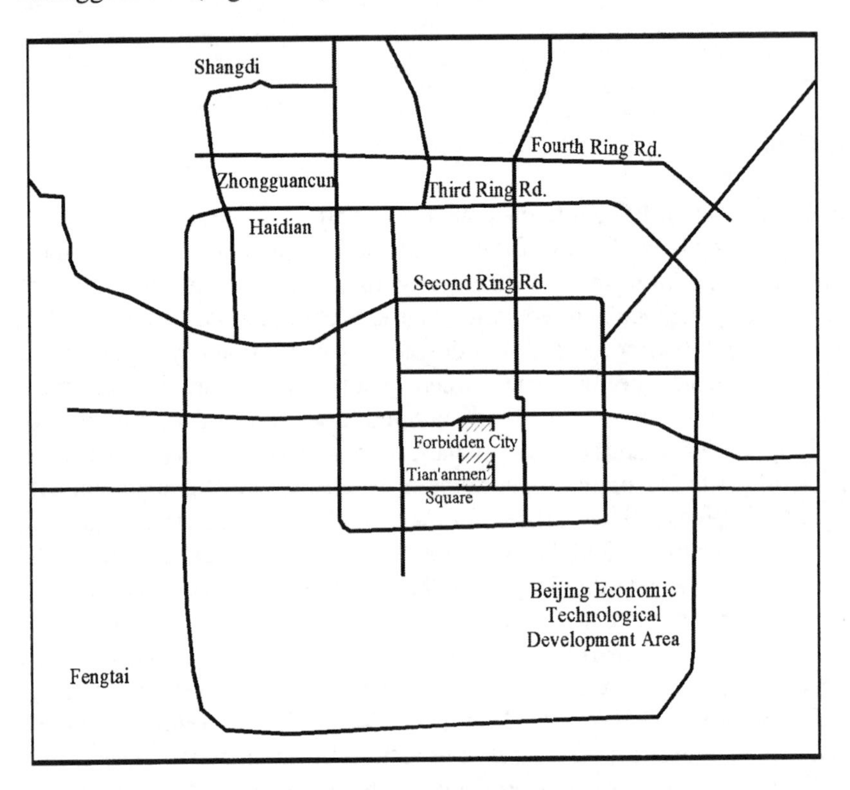

Figure 8.1 Beijing Universities and High Tech park zones

Research developments in Beijing center on the electronics industry, computer technology, nuclear industry, astronomy, and new materials. The drive to promote science and technology, with an emphasis on military applications, culminated in China's detonation of an atomic bomb in 1964 (CSSB 2000). The explosion of the Cultural Revolution from 1966-76 caused a decline in research efforts until Deng Xiaoping's reform and redirection efforts in the late 1970s. Beijing led the way to new research with entrepreneurial applications in 1980 when Chen Chunxian from the Chinese Academy of Science launched a private technology company, whose descendants became the powerful Legend Group. The company was located in the Haidan district, to the northwest of central Beijing, due to its proximity to eight of the top universities in the nation and 24 research institutes affiliated with the Chinese Academy of Science.

Computer-IT Cluster Firms

China's first cluster of High Tech related activity was created from 1984-87 when a group of computer stores on a particular thoroughfare in Zhongguancun earned that section the nickname of 'Electronics Street'. As necessary in a strongly centralized state, particularly in the capital city, the creation of something representing such a radical break with the past came about only with the approval and support of the central government. The timing was right for this development, following out of Deng's opening and reform movement. In 1984 alone, related firms in Zhongguancun mushroomed from 11 to 40 (Wang and Wang 1998). Other major conglomerates grew from individual inventions during this time period. Peking University professor Xuan Wang invention of a typesetting printer for Chinese characters became the Founders Group; Chuanzhi Liu's word processor for Chinese romanization and characters grew into the Legend Group.

The development of a high technology in Zhongguancun was formalized in 1988 with creation of the 100 sq. km. Beijing Experimental Zone for the Development of New Technology Industries (BEZ). The zone includes the subdistricts of Shangdi Information Industry Base (1.8 sq. m.), Fengtai (50 sq. km.) and Changping Park (5 sq. kilometers). The structure of separate areas with different administration within the same general area north and west of the city became officially known as 'one zone with multiple specific parks'. Nearby infrastructure connections that nurture the node include its location between the third and forth ring roads, three expressways, the biggest railway station and freight yard in China, and a local airport. In the heart of China's educational concentration, the zone encompasses 138 research institutes and 55 colleges (www.chinatorch.com). Yet another specific area, the Beijing Economic Technological Development Area of 15 sq. km., was set-aside to the south east of the main city in August 1991. Strong High Tech Development Corporation, a SOE, sponsored the establishment and management of this zone. Its major TNC tenant, the Finnish firm Nokia, spearheaded the creation of its own park-within-a-zone where its major suppliers are now located together.

The Beijing districts' successful attraction of investment in technology companies led the government to widen its geographic scope by creating 52 (as of mid-2002) additional national Science and Technology Industry Parks, so designated because they receive incentives and special attention from the central authorities. The major monetary incentive for locating in a national level park consists of significant phased tax reduction allowances. If an idea succeeds in Beijing, under the watchful eyes of central government authorities, it can then be extended to other areas of the country targeted for similar treatment. The undisputed success of Beijing's Experimental Technology Development Zone provided a launching pad for China's technology development leap into the commodification of knowledge in an increasingly capitalist system.

By 1998, slightly over 6,000 companies operated within Beijing's Experimental Zone, employing 173,000 professionals. This compares to the lower number or 50,000 people in the over 4,000 companies classified as high technology outside the BETDZ. As another comparison, more technology-intensive companies in commercially dynamic Shanghai operated outside its several distinct zones. This different spatial pattern reflects the more scattered location of Shanghai's major universities, and the municipal government's desire to further de-concentrate the intensely centralized population in the city's dense inner districts. Beijing's districts are both more innovative and more contiguous to its universities than are Shanghai's High Tech development zones, as illustrated in Type III of Chapter 2's typology models.

Beijing's major high tech industries are located along the fourth ring road circling this spread out city, resulting in a northern arc type of geographic formation. New parks planned next to the campuses of Peking University and Tsinghua University intensify the northwestern concentration. Suburbs adjoining this area attempt to catch companies looking for less expensive locations while trying to grow their own enterprises within newly established industry parks, lacking the national level status. While universities and other research institutions seek to start up their own science and technology industrial parks, new technology-intensive companies are constructing research institutes and offices outside of but dedicated to their own product development (CSTD 2000). This contrasts with the American practices of setting up in-house laboratories, contracting for outside research from universities, or merging with other companies to acquire innovations needed.

Companies headquartered in Beijing have also set up R&D facilities in other parts of China and abroad. The output of Beijing-based high technology efforts in the year 2000 is displayed in Table 8.1. Companies clearly perceive an advantage to being located in the capital city, from the associated prestige or other cluster urbanization economies. The greater value of technology exports produced outside the city indicates the location of large manufacturing plants seeking nearby but less expensive land and proximity to Tianjin's port facilities for shipping. Other disadvantages of locating within Beijing proper include traffic congestion, high land and labor costs (Cheung, 2001).

Table 8.1 Beijing high technology development

Category	Total (100 m. RMB)	Within Beijing	Outside Beijing	% Increase 1998-99
Revenue of Technology, Industry, Trade	541.34	378.64	162.7	25.2
Industry Products	348.80	232.1	111.7	27.3
Export (100 Million $US)	7.32	3.32	4.0	11.0

Source: China Statistical Yearbook (2000)

Within Haidan District's Beijing Experimental Zone, which comprises the largest area designated for high technology companies in metropolitan Beijing, the electronic information industry produces 70 per cent of the sectors represented, followed by optics, mechanics and integrated electronics (14 per cent). In 1997, Beijing's software companies comprised 20 per cent of the country's total, producing 60 per cent of the revenue nationally generated in that sector. As a proportion of the BEZ's gross production, electronic information industries again contributed the greatest share (68 per cent), followed this time by new material and new energy industries (11 per cent) and optics, mechanics and integrated electronics (ten per cent), a close third (CSSY, 2000). The leading companies in these sectors are joint ventures, reflecting the importance of foreign companies for infusing both managerial 'software' and capital 'hardware' through technology transfer from companies headquartered in more advanced countries.

Multinational Presence

In order to accelerate China's position in strategic sectors such as computer technology, the Party leadership drafted the slogan of 'go with the Giant' in the early 1990s, encouraging fledgling Chinese companies to engage in joint ventures with foreign leaders in their product area. In 1993 the three largest Chinese computer related companies gained a berth on the Hong Kong stock exchange. Sidong Corporation entered into a joint venture relationship with Japan's National Electronics and the American Compaq Corporation.

Three giants now listed on the Hong Kong Stock Exchange and allied with major global partners emerged from the early companies: the STRONG High Technology Development Group (an SOE affiliated with Japan's National Electronics and the U.S. Compaq company), the Legend Group (affiliated with U.S. Intel) and the Peking University's Founders (Fangzheng) Group, which dominates character printers and set up its own company in Japan. Former state owned enterprises (SOEs) in electronics converted in the 1990s to private companies and became partners with major foreign companies such as Siemens (Germany), Nokia (Finland), Motorola (U.S.), Panasonic (Japan), Altel (France) and LG (Korea) (CSSB 2000). Other multinational companies maintaining a presence in the Beijing Economic and Technological Area include Japan's Mitsubishi, Mitsui, National, Hitachi, NEC, Sumitomo and Fujitsu. Hewlett Packard, International Business Machines, AT&T, Microsoft, Intel, and General Electric represent North American firms.

Around 20 large multinational companies operate research divisions connected with colleges and universities in Beijing. The majority of these are located in Zhongguancun and focus on information technology development. China's first intranet connections were established in this part of the city to facilitate information exchange among related companies and colleges. Two major venture capital firms, one in Hong Kong and the other American, provide funds to fledgling Beijing technology entrepreneurs, supplementing three risk capital companies in the capital with around US$100 million in funds to invest.

Large multinational firms seeking government incentives also apply to set up special zones in which to manufacture technology intensive products. Beijing's 1,700 Taiwan-connected firms include two who created their own industrial parks within the Beijing Economic and Technological Area for the production of mobile telephones, other wireless and optical communication devices, software, and miniaturized products (Delegation of German Industry and Commerce, 2001). A new Beijing Economic-Technological Development Area started in 2000 in the southeast corner of the city, close to the airport and railroad station and stretching southeast from the fourth ring road along the capitol expressway. Multinational giants include Nokia, IBM, Sanyo, Lucent, Bayer, Coca-Cola, Cummins, BMW and Lotte, balancing the northwestern location of Beijing's university-linked homegrown ventures.

Major incentives for setting up a presence in Beijing are two-fold. Numerous corporate managers indicated that hey felt it was necessary to establish a physical presence in the seat of centralized power. Such a presence served both to make it easier for the company's concerns to be heard by increased accessibility to officials, and also for the companies to more clearly and quickly catch the latest policy promulgated from the capital. The second incentive for locating in a highly visible city with a large collegiate population is to be more attuned to modifications of product that would help a company penetrate the potentially large Chinese market as well as catch coming trends and developing brand loyalty (or at least familiarity) among the next generation of potentially affluent consumers.

The labor market ties between Chinese graduates and multinationals also develop strongly in the city of China's most selective universities. As in Shanghai, top talent often prefers to work for non-Chinese firms due to better salary and work conditions, and the opportunity to work abroad with better facilities and benefits in many life quality aspects. Given China's fairly porous intellectual property laws, much so-called technology transfer occurs when employees shift jobs or 'moonlight', both of which often work to the detriment of the initial employer. Recognizing the presence of China's largest highly trained labor pool, many companies from more developed regions advertise throughout the Beijing area for potential employees, seeking China's brightest engineers, statisticians, and a variety of scientists.

University-Research Center Model

The successful creation of companies following application of research ideas inspired Beijing authorities to set up numerous High Tech research institutes, hoping for continuing knowledge spillovers in the new entrepreneurial atmosphere of the 1990s. The success of Zhongguancun was officially acknowledged and celebrated with the 1999 report on 'The Request to Accelerate the Development of Zhongguancun Science and Technology Zone' as a model for other aspiring high technology regions, much as California's 'Silicon Valley' has achieved mythic global stature. Zhongguancun's popularity as a site for companies continues to grow (Table 8.2). The Chinese Academy of Science's Legend Group's personal computers dominate the Chinese market, rank third in Asia and fifth in global sales, sending it to the top of China's most profitable companies (CSSB, 2000). Peking University's Founders Group laser printer dominates the Asian character market.

Table 8.2 Beijing Zhongguancun science park

Category	1999	2000	2001
High Tech units	1,227	2,461	3,060
Employees	NA	293,000	361,000
Gross industrial output (billion RMB)	65.2	91.3	128.71
Profit (billion RMB)	6.75	10.15	11.3
Exports (billion US$)	96	1.82	3.05

Source: www.zgc.gov.cn (Administrative Commission of Zhongguancun Science Park

Tsinghua University's fame comes as the alma mater of numerous highly placed government officials and the primary training ground of China's engineers and scientists. It has reportedly eclipsed long-time rival Peking University as the top college choice for entering freshmen, conceivably in acknowledgement of the new paths open for scientist-entrepreneurs to wealth and prestige. In both universities, dormitory space has been set aside as incubator space for young students (and recent graduates) trying to turn ideas into profitable products. Typically, they seek to emulate Silicon Valley's birth via Stanford University, with the university providing the free overhead for hatching their business (Markoff, 2000). The success of academia-originated Legend, Founder, and Huawei high technology companies provides an inspiring example. Their fortunes rise and fall with those of their capital investors on the other side of the world – investors whose own fortunes are tied to the NASDAQ stock market.

A distinction underlining the risky nature of university spin-offs is the 'four self' model expounded for these ventures by the central government, which encouraged them at a fiscal distance. Like their models in more capitalist societies, these fledglings were to be responsible for choosing their own business partners, finding their own financing, running their own operations, and shouldering responsibility for financial losses (Wang and Wang, 1998; CSSY, 1999). The attrition rate is similarly high. Offsetting in part this 'out on a limb' instability are the various information linkages in Beijing, from area-based intranets to university-based alumni networks, occupation-based organizations (Chief Financial Officers, Chief Executive Officers), type of company (SOEs or non-SOEs, sectoral affiliation), links to formerly affiliated government departments, targeted periodicals, and web-based listserves and chat rooms. The appearance of these network-building systems indicates the vital recognition that isolation from information constitutes the greatest threat to business longevity, regardless of its age or size.

In the late 1990s, major Western companies such as Microsoft set up research and development facilities in the Zhongguancun district that includes these and several other major institutes of higher education. Almost five years later, a common complaint is that China's best brains have been hired to work for these non-Chinese companies, drawn by the relatively high salaries and benefits. Another affect could by the stifling of local innovation and entrepreneurship. However, Beijing's pool of electronics talent draws Chinese from other areas such as Taiwan, to get in on the development of the 'next wave'. The 'dot-com' bubble burst has already swamped some would-be surfers. Large companies like Microsoft see this as only a pause, as they construct their R&D lab with Chinese talent from throughout the overseas and continental diaspora.

The Peking University Economic Technology Development Area (ETDA), a 'state level' facility, contains around forty firms including 15 pharmaceutical enterprises. The sectoral emphasis of the area is designed to attract informatics and biotechnology firms. Bio-Avantis, a large U.S. joint venture biotechnology network organization, serves as the major anchor firm. A national research development institute and an extension of the PKU Medical School provide the only research and development activity; most of the other firms are representative offices of large multinationals such as Bayer, GE, and firms from Japan and Hong Kong. Billed as the beginnings of a 'Medicine Valley', the Beijing Development Area management sees itself as a service provider, furnishing infrastructure for firms to develop, such as Chinese academic institution spin-offs. The avowed model for this ETDA is Shandong University's provincial level science park, which coalesced earlier around 20 universities in the capital city of Jinan. One expression used to characterize this common practice was to 'build a nest around a phoenix, as a net to catch more attracted phoenix' (interview, Beijing High Technology Conference).

The Chinese Academy of Science (CAS) is a feature of higher education and R&D 'with Chinese characteristics' (as the common differentiating phrase goes) without a Western counterpart. This entity's Institute of Computing Technology proved particularly adept at generating and sustaining spin-offs, which additionally receive financial and other assistance from the early leader, the Legend Group. Similar to the situation in San Diego, a sort of 'incubator without walls' functions with new firms returning part of their profits to the mentoring Group but basically responsible for their own maintenance. Legend and Founders both inhabit impressive mini-campus buildings in the BEZ.

Firms born in Beijing do not necessarily stay in the area as they do in their California counterpart, however. The relatively high cost of land constitutes one consideration, in part handled by movement to surrounding suburban areas with less expensive land and more readily available labor for the low-skill manufacturing phase of the product cycle. Beijing's draw for the undocumented 'floating population', which frequently settles in outlying areas, provides a reserve labor market. The less easily remedied consideration is the undesirable proximity to central surveillance apparatus in the capital. This in part explains why up to one third of Beijing-born businesses relocate in southern Guangdong province, another booming manufacturing site though far less innovative, illustrating an internal product cycle migration familiar in the West (Markusen, 1985; Vernon, 1966).

Paradigm or Property Shift

The number of 'extensions' planned for already established science parks indicate the apparent success of these spatially configured zones, among the 15 projects listed for the Zhongguancun district alone (Table 8.3). Some of the development occurring in 2002 includes a new feature: the 'parks within a park' scheme. An example of this is the Finnish company Nokia's development of an industrial area within the new Beijing Development Area on the southeast side of Beijing. Nokia's intent is to set apart an area for its own suppliers and affiliates that will be more directly managed by the lead company, rather than interspersed with other companies throughout the BDA. By bringing a large number of new tenants, Nokia was able to negotiate more concessions and favorable terms from BDA, and inspire other major companies to consider doing similar arrangements. Negotiations are underway for other sub-STIP spatial configurations.

Table 8.3 High Tech expansion project plans, 2001-2005

Beijing Biomedicine Belt
Beijing University Science Park
Changping Science Park extension
Fengtai Science Park extension
Huilongguan Research and Development and Industrial Park
Tsinghua University Science Park
Science Park for Private High Tech Enterprises
Shangdi Information Industry Base north wing
Telecommunication and Related Product Center in Electronic City
Yizhuang Science Park extension
Zhongguancun High Tech Commercial Center
Zhongguancun High Tech Export-Oriented Industrial Park
Zhongguancun Life Science Park
Zhongguancun Science Park renovation
Zhongguancun State Software Park

Source: www.zgc.gov.cn, Administrative Commission of Zhongguancun Science Park

The type and timing of expansions in commercial areas in northwest Beijing leads to questions concerning the basis and purpose for such developments. Construction cranes along roads behind the main avenue through Zhongguancun have been quite active in the first years of the 21st century. While the main thoroughfare is lined with storefronts selling high technology products (many with non-Chinese brand names), the principal real estate type being constructed is for office space. As the cost of land along the main areas of Zhongguancun has inevitably risen to reflect the new prosperity and the marketization of land in China, start-up companies have been driven into more peripheral outlying areas. Although major universities and established giants such as Founders and Legend have also established new STIPs and expanded existing facilities in this time period, the number of market ready products has not increased correspondingly. Rather, a look at the real estate landscape reveals the increasing presence of foreign companies.

In order to evaluate the type of STIPs, incubators, and their sponsorship, as shown in Table 8.2, the different types of institutions capable of producing commodifiable innovations should be kept in mind. As discussed in Chapter 3, from 1949 through the mid-1990s, applied research in China was basically relegated to schools set up by various government departments and the various branches of the Chinese Academy of Science. These operated on funds awarded by the government as the outcome of bids to do particular projects designed and delegated by the government. The role of the universities, including top national institutions such as Peking and Tsinghua University, primarily consisted of instruction and theoretical research. In-house corporate research, common in more advanced countries such as the U.S. and Europe, did not exist. Predictably, this situation led to many frustrated professors who could not bring ideas to market, and a greatly constricted range of applied research due to the top-down nature of funding for applied projects.

Restructuring of both government and academic arenas since the late 1990s dramatically changed this picture. The consolidation of government departments led to the phasing out of their associated research institutions and state owned enterprises. Universities were and continue to be encouraged to contribute to developing innovative (or adapted) products, and are now bidding in competition with the Chinese Academy of Science for project contracts and grants. Many university students have caught the highly infectious 'Silicon Valley fever', choosing to spend their time trying to develop new products rather than mastering or extending basic principles that science relies on for major breakthroughs. Following State mandates, incubators have sprung up throughout Beijing since 1988 with a variety of sponsorships to nurture new ventures (Table 8.4).

Table 8.4 Beijing area incubators

Name	Type
Biology City, Peking University	Peking University
Biological Medicine Research Center	Peking Medical University
Science & Technology Park	Tsinghua University
Software Company	Tsinghua University
Research Center of New Materials	Beijing Aviation & Astronauts University
Beijing Experimental Zone	STIP incubator
Beijing High Tech Creativity Service Center	STIP incubator
Shangdi Returned Student Enterprise Center	State owned STIP incubator
Fengtai International Enterprise Center	STIP incubator
Changping International Enterprise Center	STIP incubator
Daxing County	Private ownership
Zhaoyang District	Private ownership

Source: China Science and Technology Department (1999)

Tsinghua University's 'Pioneer Park', with occupants since 1999, preceded Peking University's incubator that is still under construction (Harwit, 2002). Occupants all received a Tsinghua education, and benefit from retained ties to graduate professors and their laboratories. Particular efforts revolve around soliciting graduates to return from overseas, partly due to provisions that they can bring a large amount of their own capital to invest (US$36,000 - US$48,000) and partly due to their experience in overseas firms, representing an otherwise unavailable source of practical knowledge. Similar programs exist at major Chinese universities profiled in other chapters, including Xi'an's Jiaotong University and Shanghai's Fudan University. Drawing on alumni ties, these university incubators can sometimes draw on highly sought after foreign venture capital, particularly from more established and venturesome overseas Chinese who seek to make a contribution to the university – and get in on the ground floor of a potential company with high returns not available in less risky environments. These business investors are also more accustomed to the customary high failure rate of new enterprises than are Chinese banking and government lenders who prefer more secure investments.

Sluggish SOEs also seek to breath new ideas into their moribund existence by installing in-house incubators for related start-ups. Although only 15 existed in SOEs in mid-2002, they were concentrated in key sectors such as pharmaceuticals and biotech companies, chemicals, iron and steel, and engine plants (Harwin, 2002). Although advice in much-needed management tactics was purportedly available, the quality and frequency of this assistance is spotty. Higher salaries and benefits are available elsewhere for these high-demand talents.

Zhongguancun area high technology enterprises can be linked through China's first intranet, followed by one constructed by Tsinghua University (China Colleges and Universities Science and Technology Corporation Network) and several others set up under public-private auspices. But connectivity in the physical sense, while enhanced by proximity to Beijing's major research center resources, does not guarantee creativity. Numerous, and potentially serious, questions remain that touch on the institutional nature of systemic 'Chinese characteristics'.

Challenges

Access to adequate capital remains a problem, shared by fledgling companies in the developed world but exacerbated by the shortage of private investment capital in a recently marketized country such as China. The national government provides physical infrastructure such as building space, desks, computers and printers, seed capital from programs such as Torch, access to the technological business structure and a legal framework, all necessities and improvements on the past that put Chinese start-ups on par with global peers. Experienced advice provided by business service personnel, from seasoned managers to accountants, lawyers, and financial experts, is in short supply nationally as well as in technopoles like Beijing.

BEZ management functions to provide some advice and assistance in areas such as loan application, tax relief, and even entrepreneur image enhancement (Gu, 1996). Newspaper articles and television news feature services provided by both foreign (often expensive American firms) and Chinese (more affordable, less experienced) business consultants. A positive spin given successful ventures has successfully created an acceptance of and aspiration toward becoming an entrepreneur. The number of high tech companies in zones such as the BEZ exhibits steady increase since establishment of the zone.

An overly enthusiastic, somewhat infectious attitude playing off the Nike slogan of 'Just Do It!' grips enough inexperienced entrepreneurs to cause concern, however (Wang and Wang, 1998, personal interviews). Given the shortage of government funds in general and for risk ventures in particular, the traditional Chinese technique for raising investment capital consists of pooling donations from related individuals, relying on kinship, friendship, or even college class ties in the classic exercise of *guanxi* (connection). The failure rate of such ventures is difficult to assess, but due to its nature carries more social and personal consequences, particularly in a society characterized by high savings rates, few but deep demands for cash (weddings, emigration, illness), and general lack of disposable income. The outreach to deep pocket multinationals for financing, through joint ventures or mergers and acquisitions as is prevalent in the U.S., co-opts the local benefits of indigenous innovative R&D while providing incentive for invasion by the global giants.

Isolation is the enemy of innovation. Without communication, and application after testing, ideas perish. Networks are needed, particularly in newly innovative and entrepreneurial developing world settings, so location in a park would seem to be ideal. The degree of interaction in a park setting can, however, be fairly restricted on the one hand, or overdone by bureaucratic micromanagement, on the other.

The challenges faced by entrepreneurial high technologynology firms in the West are repeated and compounded under developing world and transitioning market conditions in China. While active and on-going efforts by government and research institutions to provide a shelter in high tech parks can partially assuage the situation, much like typhoon shelters for small boats, the storm of competitive pressures still rages.

China's odds of building its own learning districts as contributing and desirable nodes in a global network, rather than merely as sites for domestic market penetration or skilled labor pools, can only be enhanced through establishment of science and technology industry parks in throughout the capital.

Chinese companies experience a debilitating lack of venture capital, but to a different degree than in the West. In China, the central government furnishes the majority of venture capital. With this comes a clear responsibility to spend the 'people's money' carefully, imposing a brake on risk investment. The source of funds in Beijing (Figure 8.2) is more government oriented than in the rest of the country (Figure 8.3), where proximity to the government lessens and is thus less attractive. Investors always desire to be closer to their investments, so proximity plays a key role. In both cases, banks play a much smaller role in financing S&T, given their problems with liquidity in general and their role in providing as a government-mandated source for its own major projects, from constructing new infrastructure to sustaining old and ailing SOEs.

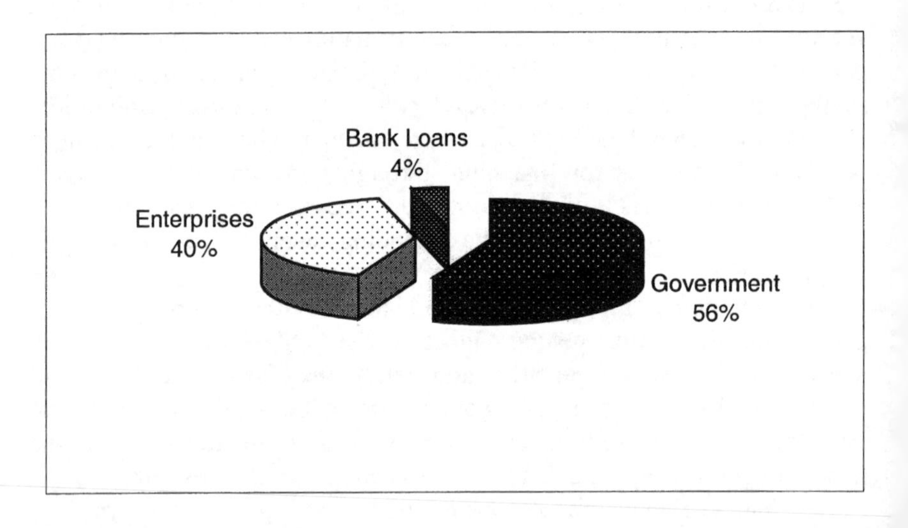

Figure 8.2 Sources of Beijing's science and technology funds

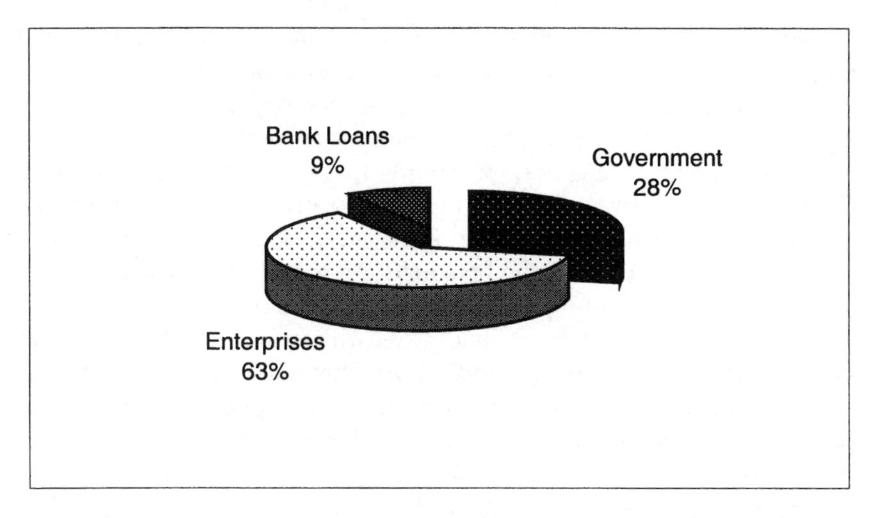

Figure 8.3 Sources of science and technology funds nationwide

Beijing venture capital funds include both government-supported entities and joint ventures with foreign funders, particularly American partners. The Beijing municipal government itself assisted in the creation of a fund with Hong Kong investors and an American partner (American Zhongjing Group). Chinese banks have created 'special credit and loan projects' with well-known established high technology firms such as Legend and Founders to extend less risky credit for business expansion. The strong involvement of government in Beijing's new businesses, however well intentioned, has been seen as retarding their development by suppressing the need and development of mechanisms for more market-based funding and ownership structures. Enforcement of transparency in legal codes such as accounting practices and employee stock ownership could provide incentives for investment to raise capital as well as productivity, as they do in more advanced countries (Cao, 2001).

The proliferation of development districts, STIPs, and firms within them illustrates the principle predicted virtues of proximity, especially in a labor pool of young, well-educated workers in high demand. These include a high turnover rate of employees at high technology companies, taking their technology knowledge learned on the job as well as in the university with them, and the strong ties among students in similar graduating classes. Lax enforcement of intellectual property laws feed this situation, to the chagrin of foreign companies who suffer the most from such 'informal' technology transfers.

While proximity to China's foremost research institutions draws high technology companies, proximity to the strong supervision of China's center of government has the opposite effect. As discussed in the chapter on Shenzhen, this Pearl River delta metropolis on China's southeast coast by Hong Kong benefits from numerous ties with Beijing based companies and universities. As long as there remains an 'emperor', or strong 'imperial' presence, innovators will seek the higher and freer skies elsewhere (*tian gao, huangdi yuan*). Other large cities with formal bureaucratic structures such as Shanghai also inspire entrepreneurs to flee to their outskirts (such as Suzhou) or beyond, seek more flexibility in response to the new and often little understood and even less accommodated needs of high technology entrepreneurs. The key question for the future remains whether Beijing's spatially clustered companies will develop into true innovative entities, or degenerate into weak nodes exploited by a global network of multinational, externally headquartered giant corporations.

References

Cao, Cong (2001), 'Zhongguancun: China's Silicon Valley', *The China Business Review*, vol. 28, pp. 38-41.

Cheung, C. (2001), 'The Zhongguancun Industrial Cluster in Beijing, China: Its Actors and Their Roles in the Industrial District', Paper presented at High Technology Conference, Beijing, PRC.

China Science and Technology Department (2000), 'Report on High Tech Industry Development in Beijing', *Development Report on China's New & High Tech Industry*, China Science Publishing Department, Beijing.

Delegation of German Industry & Commerce (2001), *ChinaInfoFlash*, www.ahksha.com.cn, 12/2001.

Gilley, B. (1999), 'Looking Homeward', *Far Eastern Economic Review*, vol. 162, pp. 50-52.

Gu, S. (1996), 'The Emergence of New Technology Enterprises in China: A Study of Endogenous Capability Building Via Restructuring', *The Journal of Development Studies*, vol. 32, pp. 475-501.

Harwit, E. (2002) 'High Technology Incubators: Fuel for China's New Entrepreneurship?' *The China Business Review*, vol. 29.

Hotz-Hart, B. (2000), 'Innovation Networks, Regions, and Globalization', Clark, G., Feldman, M. and Gertler, M. (eds.), *The Oxford Handbook of Economic Geography*, Oxford University Press, Oxford, pp. 432-54.

Markoff, J. (8/4/2000), 'Silicon Valley's Primal Spirit Lives On, in a Part of Beijing', *The New York Times*, pp. A-1,4.

Markusen, A. (1985), *Profit Cycles, Oligopoly and Regional Development*, The MIT Press, Cambridge, MA.

Park, S. O. and Markusen, A. (1995), 'Generalizing New Industrial Districts: A Theoretical Agenda and an Application From a Non-Western Economy', *Environment and Planning A*, vol. 27, pp. 81-104.

Vernon, R. (1966), 'International Investment and International Trade in the Product Cycle' *Quarterly Journal of Economics,* vol. 80, pp. 190-201.

Wang, J.C (2000), *Chinese Industrial Clusters*, Beijing University Press.

Wang, J.C. (1999), 'In Search of Innovativeness: The Case of Zhongguancun', in Malecki, E. and Oinas, P. (eds.), *Making Connections: Technological Learning and Regional Economic Change*, pp. 205-30, Ashgate, Aldershot, UK.

Wang, J.C. and Wang, J.X. (1998), 'An Analysis of New-Tech Agglomeration in Beijing: A New Industrial District in the Making?' *Environment and Planning A.*, vol. 30, pp. 681-701.

Wang, S., Woo, Y. and Li, Y. (1998), *Development of Technopoles in China.* Asia Pacific Viewpoint, vol. 39, pp. 281-301.

www.chinatorch.com/stipark/english/page91.htm, (2000), 'Zhongguancun Science and Technology Industrial Park'.

Chapter 9

Local Innovation Learning Zone: Shenzhen

The North-South Axis

Shenzhen's position as a 'local innovation learning zone' is quite distinct from that of previously profiled Beijing, which as the capital city benefits from the presence of China's most prestigious, long-established research institutions. The very newness of the 'Shenzhen Science and Technology Industrial Park', established on largely rural agricultural land like Hong Kong a century before, posed problems for an area avowedly dedicated to technology development: it lacks anything like a premier research university to serve as the innovation generating institution. Shenzhen municipal and industrial park authorities attempted to remedy this situation by luring a number of faculty from Beijing based universities (particularly Tsinghua) to the ostensibly sunnier, warmer, less congested and less polluted setting of Shenzhen University. Other avenues to provide access to top level research institutions and laboratories are under construction, but the need is still felt.

The impetus to develop Shenzhen as a model of high technology growth in southern China, as well as a low-cost labor, manufacturing center, springs from the post-Mao geographic strategy of allowing the coastal cities to develop first, as well as the Maoist era urge to balance growth throughout a broad area (Weng, 1998; Wei, 1999). Major cities were to be strengthened to act as growth poles for surrounding areas, reducing over time the rural-urban divide by providing jobs and spreading prosperity. In a developing country with limited resources such as China, such after-the-fact descriptive models work less well than in the West. The need to concentrate scarce resources in order to bridge the infrastructure and amenity gap with more developed countries leads to clear spatial divisions in development zones. This can be seen at every spatial scale, from the national (Special Economic Zones), provincial (Open Cities), municipal (industrial parks) and even within park zones, as is the case with SHIP's location surrounded by the larger Shenzhen Industrial Park. Inevitably, some resources are less transferable than others. Shenzhen served as China's first experiment with a modern, designed new city in the earliest stage of opening and reform. Tutelage and strength was to come from Hong Kong, directly across the river from this instant metropolis. The enthusiasm of businesses from more developed countries for locating in China waned with experience of the difficulties during the very gradual transition.

Shenzhen's difficulties with attracting and anchoring multinational companies have been well documented (Wu, 1999). Some of the problems were part of a learning curve (few skilled engineers or managers, no industrial base for supplying local capital goods or business services), others were institutional in nature (lax intellectual property protection, domestic immigration restrictions), while the more challenging and persistent are in the nature of the place (smaller and less developed than Beijing or Shanghai, a populace unwelcoming to non-Guangdong speakers, no major research centers, less developed links with inner China than with Hong Kong so reliant on imported rather than domestic inputs). Shenzhen's reputation thus is being built more now by its attraction of native entrepreneurs, particularly those coming to Shenzhen High Tech Industrial Park (SHIP). However, a continuing challenge lies in the need for a High Tech area for a major research institution capable of generating necessary knowledge spillovers that manufacturers can then commodify. An examination follows of this operation, several examples of relative success, and a look at their roots.

Shenzhen Hi Tech Park

University-affiliated entities (UAE) constitute a distinct form of high technology companies. Top Beijing-based universities directly involved in nurturing SHIP companies include Peking University and Tsinghua University. Major SHIP Chinese companies specializing in computers, telecommunications, and biopharmaceuticals maintain R&D relations with firms in more developed countries such as the U.S., and draw upon particular universities listed as participants in the 'Virtual University'. Over 33 campuses in China and abroad are part of a 'Virtual Campus' link, attempting to compensate for the lack of a strong local university. The 'Virtual Campus' (Figure 9.1) consists of a set of computers, each dedicated to a particular affiliated education institution, through whose portals prospective students could potentially take classes. Shenzhen University's strengths do not reportedly lie in technical areas. Ties to northern universities such as Peking University and Tsinghua University include incubators and extension classes, some of which are already in place and others are being set up. Several start-up companies profiled in this chapter are in various facilities maintained by Peking University, while Tsinghua is directly involved in running an attractive new multipurpose incubator and office compound.

Figure 9.1 Shenzhen Hi Tech Park 'Virtual University'

Research and development links are also maintained with nearby universities in Hong Kong. A joint MBA program offered by Peking University and Hong Kong Technical University seeks to meet the needs of local companies. The two-year course provides a large profit stream in meeting a substantial student and corporate demand for more locally available market-targeted training.

Park management tries to provide some of the best practices available in high technology clusters in more advanced countries. University lectures are advertised to companies, and available on the internet-connected 'Virtual University'. Managers are aware that companies do best when they are exposed to fresh ideas, practices are questioned, information exchanged, and interactions between researchers, practitioners, and businesses are available (interview and lecture, 7/2000). However, these arrangements are often left for interested organizations such as universities to arrange for their own alumni. Park management arranged no other sectoral or occupational networks, unlike in the United States. SHIP benefits from the lack of allocation limits on university graduates, and its positioning as a 'Mandarin speaking' entity where educated individuals throughout China should feel welcome and on common ground. Some interviewees indicated they preferred locating within SHIP due to its 'prosperous atmosphere' and the availability of university professors for consultation on business problems. Although the available labor pool usually did not include graduates from 'the best universities', employers generally felt that labor was quite adequate, hard working, and paid at wages that permitted a profit for the company (interviews, 7/2000). The declared profit margin of Shenzhen's 'new and High Tech enterprises' is shown in Table 9.1. Although the large jump in 1997 profits is balanced by the dip in 1998, profits are still above those in 1996 and overall show a steady, if undramatic, increase.

Table 9.1 Pre-tax profit of Shenzhen High Tech enterprises
(in 10,000 yuan per employee)

Year:	1993	1994	1995	1996	1997	1998
Profit:	4.89	5.66	5.62	6.96	10.06	7.52

Source: China Science and Technology Department, 2000

The role of private Chinese-owned enterprises in contributing to this amount is of some pride to the authorities. Their numbers and value of products increased dramatically in the 1990s. In telecommunications, Huawei Corporation dominates more then 25 per cent of the national market. The 'Huawei Management Pattern' is held up as a model for

High Tech companies (computer, telecommunication, measuring instruments, biotechnology, and microelectronics) nationally (China Science and Technology Department, 2000). However, the over 200 colleges, universities, and scientific research institutes cited as cooperating with Shenzhen enterprises nevertheless leave an unaddressed need that Beijing's top universities are involved in filling.

Role of Peking University

Much of the non-governmental capital invested through Peking University comes from the Founder's Group. This conglomerate originated with the outstanding success of an electronic publishing system for printing Chinese characters, now used on a global scale, for transforming electronic format to a print media. Prof. Wang Xuan, the scientist-founder, proved to be both a technical and entrepreneurial genius. His willingness (in the manner of his American computer tycoon counterparts) to invest in other promising tech start-ups magnified the impacts of his company's success. The hand of Founders extends from golf courses in Chengdu to a science and technology park in Dongguan, a later case study.

Other university corporate connections serve business needs not elsewhere addressed. Legend, the leading Chinese domestic computer brand, also funnels profits through Peking University since its founding professor was affiliated with the university. The Peking University-connected Leader Group provides services such as advice to companies in exchange for re-routed corporate taxes owed to the government. The national government directly provides land at low cost to companies with an approved high tech project, screened by the Chinese Academy of Science for its likelihood of producing a commercial product.

While the national government serves as a major capital source, the problem is how to choose the best recipients. Funneling funds through universities such as Peking University is one route (Wang, 2000). Several Chinese universities have cooperative arrangements with foreign MBA universities, such as Wuhan with Toronto and Fudan (Shanghai) with MIT's Sloan. Founder funds pass through Peking University's Research Group, which also transfers research university-affiliated ventures. Corporate profits in part then flow back to the university through the Peking University Education Foundation. The Founders Group currently sponsors more than 20 companies. A mediating organization such as Founders shortens the learning curve for fledgling ventures.

Survey and interview evidence point to the prominence of venture capital, largely through investors outside China such as from Hong Kong (outside China's economic system constraints), Taiwan, and the United States. Venture capitalists tend to provide checks, but little management advice and involvement despite the urgent and apparent need for such 'software'. This deficit forms a critical distinction from practices inside the United States, for example, where management information is more easily obtained. Universities handle technology transfer for their own students, professors and alumni. In some cases students are allowed to take a year off from classes to experiment with setting up and running a company within the supportive confines of their university-run business incubator.

Role of Tsinghua University

Incubators at SHIP furnish needed subsidized surroundings for fledgling companies. Tsinghua University recently opened its own facility in striking new surroundings designed by its School of Architecture. Before 1995, Tsinghua personnel primarily conducted R&D in cooperation with universities in the northern provinces, closer to its main Beijing campus. At the behest of the governor of Guangdong Province, a Tsinghua University alumna, the university sent a professor and money to set up a fund to support some promising (to be determined) industry in the area. Instead, the university decided that Shenzhen (whose mayor is also a Tsinghua University graduate) was a good place for a direct cooperative public-private venture. Setting up this Institute complied with SHIP's requirement that technology developed in Shenzhen should be quickly useable by industry.

The Institute serves multiple purposes. These include providing technological solutions as a service to factories in the surrounding SHIP. The incubator Institutes also supplies rentable laboratories for electronics design automation and network applications, optical mechanics, and new materials research. Some of this research directly applies to government initiatives for opening up other areas of China for development, such as investigating different types of asphalt for roads and airport runways that can function under extreme and varied climatic conditions typical of China's western provinces. Automation design laboratory facilities function in cooperation with a Silicon Valley laboratory started by Tsinghua University graduates 15 years ago. This facility currently transfers technology ideas back to China via production orders. While equipment is largely imported, most of the materials involved in production come from China. Over 300 people occupy the incubator at present, with an average college degree higher than the master's level.

Quality of life and information exchange enhancing features abound. Several coffee shops and cafeterias provide places to mingle, get to know other incubator tenants, build relationships and exchange information. The small bookstore on the premises contained primarily pirated or translated copies of books on technical (computer, IT) and business subjects. Most employees of incubator companies reportedly live in Shenzhen; amenity housing is available within a convenient ten-minute walk. Future plans for the facility include housing for second stage companies (incubator graduates), and developing an Information Park in the north district of SHIP.

Local Firms: Case Studies

SHIP companies affiliated with Peking University include Kexing (a bio-pharmaceutical firm), Jadebird (a prominent IT company with a coveted listing on the Chinese stock exchange), Vanguard (a start-up venture specializing in a proprietary innovative fingerprint ID technology much in demand by government agencies in particular), and Hong Kong's Ing Hai Wei company (a telecommunications firm). The first three firms are discussed in more detail below.

Kexing Bio-pharmaceutical Company

A member of Peking University's Weiming Group (named for the lake in the old part of the campus which doesn't have a name (*wei ming*), Kexing pharmaceuticals exemplifies companies assisted by Peking University's 'Founder's Group' investment arm. Kexing is the largest genome product company in China (Weiming, 2000). Originally a joint venture with Singapore, a parting of the ways came over marketing strategy differences. Peking University stepped in to buy out Singapore's stock and absorb Kexing as a research institute. Peking University's 'Industrial Department' supervises solicitation of venture capital, going to Shanghai among other places to learn how to raise the scarce but vital finance. One of Kexing's key breakthrough's came with the understanding that technical and management problems are linked through the production process – another step in learning the complex workings of a capitalist market-oriented venture.

Kexing is currently adding high quality manufacturing facilities to expand into an anchor of the anticipated SHIP 'Medical Valley'. A major new investor is a U.S. company, seeking to ensure the addition will meet FDA standards for export to global markets. Since the government limits foreign medicine in the genome realm, major competitors are other Chinese companies. Advertising is also strictly limited, so companies must rely on a national network of doctors and greater familiarity with the Chinese market. Given the long time it takes to develop a biotech breakthrough into a market-ready product, Kexing purchases U.S. biotechnology companies to acquire their expertise.

Jadebird Computer Technology

The manager of Jadebird, officially a state-owned enterprise due largely to the sensitive and proprietary nature of its products, graduated from Peking University along with many of the company's middle managers. It was a good year for transitioning aspiring intellectual-capitalist entrepreneurs. The company started at the end of 1994, with an investment of 200,000 RMB, including funds from the Peking University Department of Computer Science. In 1999, capital on had reportedly topped 50 billion RMB. Problems predictably arise with the university over whether the company or the university owns what per cent of corporate profits and inventions – difficulties also familiar to their counterparts in developed countries.

The basic product featured by Jadebird consists of system integration networking of computers and servers, as well as a cable TV network. Jadebird plans to join the Hong Kong stock market, aspiring to ascend to the U.S. NASDAQ listing in the following year. Another key product is a standardized chip, used by global positioning systems to track traffic flow in a GIS database including large metropolitan taxi companies. International connections include going to India to learn how to do business in computers and computer software. China acknowledges a software lag with its Asian neighbor. India's experience with the U.S. market, and ability to realize a quick profit using software

Figure 9.2 Incubating low-cost local hopefuls

products, inspires much admiration among Chinese computer entrepreneurs.

Fingerprint ID

On a different scale, a resident of the SHIP incubator – an aging building largely lacking amenities such as air conditioning in this steamy southern climate – functions as a Peking University joint venture just beginning to set up shop on its own. The basic product is an identity card with an embedded computer chip. The proprietary chip records the cardholder's unique fingerprint, which is passed through a device to check the print with previously recorded permitted prints. Current users of this technology occupy opposite ends of the demographic divide. Schools in the highly competitive Chinese system test the identification of test takers, and Shenzhen's Social Security system checks the identity of recipients of State benefits. The major stockholder and intellectual property owner is the Founders Group. The government of Malaysia also seeks to set up a joint venture with the promising young company.

Domestic Networks

Approximately 10,000 graduate students find employment in SHIP, which promotes the prominence of Mandarin (as opposed to the local Cantonese dialect) with a sign at the Park entrance, advertising linguistic compatibility for nationwide migrant tech workers. The average age in SHIP is under 30 years of age. The 'human allocation', or migration permit, for technical workers is reportedly unlimited. Police inspection patrols board buses to examine passes for permission to be where passengers are going, however. The perceived dearth of culture in Shenzhen complicates matters for companies competing for workers, technicians, and managers with establishments in Beijing and Shanghai. Costs to attract employees rise to compensate for such shortcomings, driving low wage production roles into outlying areas like Dongguan, between Shenzhen and provincial capital Guangzhou (Sit and Yang, 1998). Migrants from the countryside and inner China come to Dongguan for jobs in basic assembly operations (see Chapter 5).

Information exchange in SHIP arises in part from frequent employee churning, averaging job changes every two or three years. Quality of life amenities also provide convenient mixing opportunities, such as a large cafeteria in the basement of the nearby administrative building. Human networks affiliated with Peking University include groups of students studying hi-tech parks, from Shenzhen to Xi'an, for a summer project. Alumni linked through past service as elected representatives to the Student Union form webs of reference and assistance beyond graduation.

In addition, Peking University and Beijing's Tsinghua University are establishing a 'University Park' and enlarging already invested firms. Other primary R&D links are with Hong Kong universities. A joint MBA program offered by Peking University and Hong Kong Technical University seeks to meet the needs of local companies. The two-year course provides a profit stream while meeting a large student and corporate demand. Other universities considering setting up branches in the Information Park reputedly include Nanjing, Harbin, Fudan,

Electrical University, and some foreign universities, such as Hong Kong University and the Chinese University of Hong Kong.

Technology transfer infusions come from personal ties to university professors in the U.S. who come to an affiliated company for a month annually, student exchanges with Chinese universities, and consultations with their local professors. Most companies concentrate their efforts on manufacturing, with primary R&D done elsewhere. The participation of industry-focused university outreach signals future potential, as China encourages the development of endogenous innovative industries capable of competing at all levels within the global economy (Gu, 1996). Lacking a major university or research center within Shenzhen invariably cripples the effort to make this early leading development zone a true location for innovation. It's fusion of functions, from TNCs to incubators and local companies with overseas extensions, places it in the second category of 'District Hub and Spoke' discussed in Chapter 2. SHIP is the story, drawing less on its local connections than its own image, and sponsors (largely from Beijing) that seek to help it grow as a Pearl River delta balance to other areas around China.

References

China Science and Technology Department (2000), 'Report on High Tech Industry Development in Beijing', *Development Report on China's New & High Tech Industry*, China Science Publishing Department, Beijing.

Gu, S. (1996), 'The Emergence of New Technology Enterprises in China: A Study of Endogenous Capability Building Via Restructuring', *Journal of Development Studies*, vol. 32, pp. 475-505.

Sit, V.F.S. and Yang, C. (1998), 'Foreign-Investment-Induced Exo-Urbanization in the Pearl River Delta, China', *Urban Studies*, vol. 34, pp. 647-77.

Wang, J.C. (2000), *Chinese Industrial Clusters*, Beijing University Press, Beijing.

Wei, Y. D. (1999), *Regional Development in China: States, Globalization, and Inequality*, Routledge, London.

Weiming (2000), *China Peking University Weiming Biotechnology Group*, Weiming, Beijing.

Weng, Q. (1998), 'Local Impacts of the Post-Mao Development Strategy: The Case of the Zhujiang Delta, Southern China', *International Journal of Urban and Regional Research*, vol. 22, pp. 425-42.

Wu, W. (1999), *Pioneering Economic Reform in China's Special Economic Zones: The Promotion of Foreign Investment and Technology Transfer in Shenzhen*, Ashgate, Aldershot.

Chapter 10

Local Innovation Learning Zone: Xi'an

Xi'an, which means 'Western Peace', occupies a historically important and strategic location in central western China. This region is also referred to as 'inner China', to distinguish those places distinctly inland from the booming eastern coastal cities. The present city is proud of its two thousand-year-old intact walls. Completely surrounding the old urban core, the walls now stand largely as they were restored in the Ming Dynasty (AD 1328-1644, the last ruled by a native Chinese ethnic Han). Not far away lies Banpo, where the oldest relics of Chinese civilization were unearthed on the banks of a tributary of the Yellow River. The tomb of fabled and feared emperor Qin Shi Huang (BC 259-210), brutally effective if short-term unifier of the Chinese state and first contractor for the Great Wall, also lies close by. As Chang'an, or 'Long Peace', the city served as the capital of the Han dynasty (BC 202), the first major kingdom to adopt Confucianism and anchor the Chinese end of the 'Silk Road'. This early trade affiliation consequently gives the city a Muslim flavor from its large Hui ethnic minority. Xi'an's relatively remote location by the mid-20th century commended it to Mao Zedong, who followed Stalin's World War II model by defensively moving development capacity of a militarily strategic nature to an inland location.

The city currently covers an area of 10,108 sq. km. including eight districts and five counties, with a population of 6.9 million. Out of these, four million people live in the downtown area of 180 sq. km. (Administrative Committee, 2001). Typically for China, males outnumber females 3,450,00 to 3,232,000, with a gradual three per cent annual increase in 2001. Xi'an serves as the largest transportation, telecommunication, information and finance center in China's equivalent of the American Midwest. Relative to other major cities, Xi'an lies an official nine hours by road west of Beijing, and 15 hours by road northwest of Shanghai.

Silk Road to Third Front

Science and technology industrial parks (STIPs) in west central China's ancient capital city flourish with far fewer multinational corporations than their counterparts contain in coastal China. The major local engineering institute of higher education is Jiaotong University, which moved its main campus from Shanghai to Xi'an in 1956 as directed by the central government. The main motivation for the transfer was to support military enterprises, especially missile development, in the face of what was perceived as a threat from U.S. support of Chiang Kai-shek's Taiwan-based troops. Given its roots in the government's military-industrial policy known as the Third Front (Naughton 1988) from 1964-1972, and the current push to develop inner China to balance the success of the east coast, not surprisingly many of Xi'an's major industries reflect strong ties to the central government. They serve as the exemplar of Markusen's 'state-anchored industrial districts' (Markusen, 1994).

Characteristics applicable to Xi'an's development structure from Markusen's model include:

- Domination of the business structure by large government institutions and key investment decisions made at various levels of the government.
- Labor market relatively fixed to place rather than business due to a unique local cultural identity and low drawing power from outside area.
- Little 'patient capital' (long term non-risk averse) available inside the area.
- A high degree of government involvement in infrastructure, information provision, trade associations, and future prospects for growth.

Government Model: Military-Industrial Complex

Launched in the late 1950s, Mao's Third Front movement sought to create research and military-industrial centers within China's interior as alternatives to east coastal centers that developed over the previous century in response to contact with the outside more advanced powerful nations. China's Russian mentor in this time period looked back on their World War II development of areas east of the Ural Mountains as an effectively safer geographic defense during Hitler's invasion of their more advanced western region. China's intention lay in diminishing its vulnerability to air strikes by the American military seeking to quickly knock out the fighting capacity of an enemy of their ally Taiwan. Leaders in Shaanxi seized this opportunity, and those presented from 1964 to 1972 during the Great Leap Forward period, to augment the economic strength of their province's investments in transportation and communication infrastructure, as well as military-oriented research and development projects and production (Bachman, 2001; Naughton, 1988). The province's infrastructure now includes a railroad connection to Nanjing, an airport due for upgrades as befitting its stature as the gateway to the (projected) booming Western China, and optic fiber networks furnished by Canada's Nortel corporation.

President Jiang Zemin used Xi'an as the site for announcing the central government's millennial 'Great Western Development' campaign in June 1999. It was promptly dubbed by some Westerns 'Go West, Young Han' to highlight the unspoken objective of increasing the proportion of Han migrants to a region of China in which they are less prominent. The plan highlights ten infrastructure and basic resource development projects, in addition to a software development project with IBM at the High Tech end investing in a platform development project (CNWA 2000). Fiscal incentives include five to eight years of graduated tax holidays, tariff-free incentives for specified imports needed for manufacturing not available in the quality and quantity required in China. Infrastructure development concentrates on connecting Xi'an to other development areas via a new railway to Nanjing and a projected new international airport. Three belts along major transportation routes are envisioned as future linked development zones in the West:

- the railway between Urumqi, Lanzhou and Xi'an.
- the Yangtze River area in Sichuan and Chongqing provinces.

- the Nanning-Kunming Railway running through Guangxi and Yunnan provinces.

Xi'an's four industrial parks, in different sections of the city, minimize commutes to around ten minutes (Figure 10.1). Of these, three are State level developments out of the 53 'National High and Technology Industry Development Zones' (www.chinatorch.com, 2000).

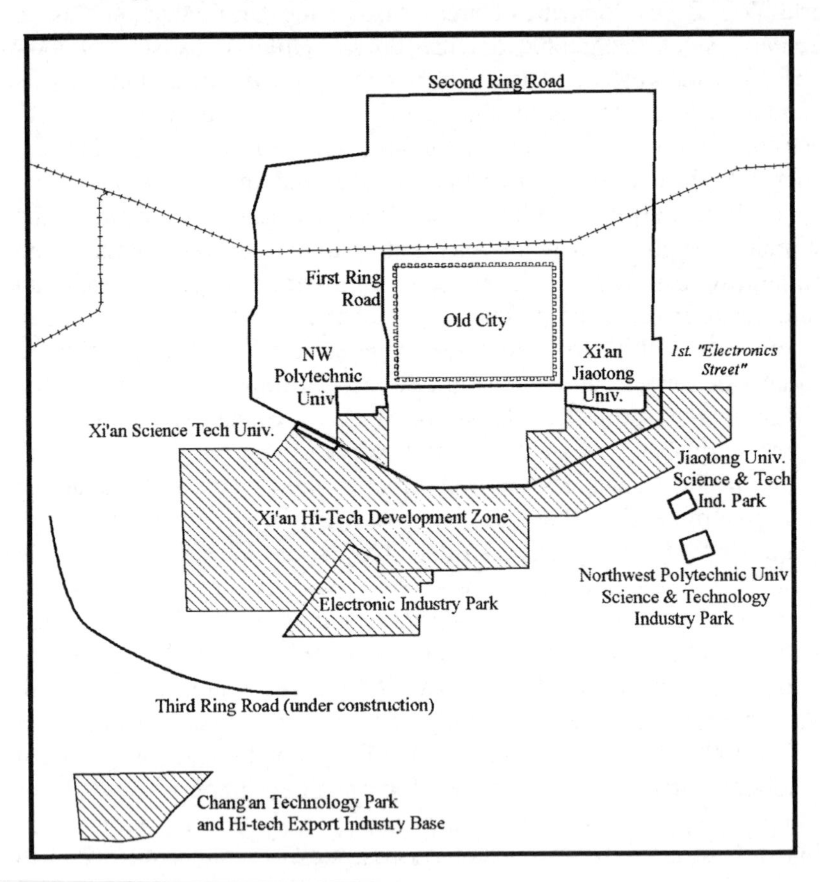

Figure 10.1 Map of Xi'an

Each park is staffed with a government representative to help resident companies make connections with various government branches such as tax, law, and customs in order to stay informed of and compliant with the latest regulations. These regulations continue to favor both

Chinese and foreign investments for development in the West, despite WTO regulations barring special incentives for foreign investors in any of its cooperating member nations. In its current efforts to redistribute the wealth accumulated in the coastal provinces since 1978, and despite the windfall in tax revenues foregone by such provisions of local enticement, China's central government continues to geographically advantage certain of its regions. These steps are needed but perhaps insufficient, given the huge difference between the extensively developed infrastructure and core of FDI companies on the eastern coast.

The Xi'an HTIDZ, established in 1987, is one of five 'Key High Technology Zones' designated by the Ministry of Science and Technology for support by the national government during the next five to ten years as part of the Five Year plans. Outside of established government funding programs such as the Torch initiative for start-up businesses, Singapore and Japanese investors are the main source of investment capital, which is in generally short supply. Xi'an's capacity to absorb the amount of money flowing in falls short of the pledged amount. Of US$595 million contracted at a major trade fair in late 2000, only US$156 million was utilized as foreign direct investment in the following year (Economist 2002). Primary funders of park enterprises include national, provincial and municipal government sources (www.xjtu.edu.cn/xjnet/industry 2000).

The primary sectoral foci of enterprises in the Xi'an High-Tech Industry Development Zone (HTIDZ) are electronic information, electro-mechanical integration, refrigeration technology, and biomedical pursuits. Of 411 foreign invested enterprises in the zone, the majority of whom receive funding from Japan, the U.S. and Germany (respectively) as well as 25 other countries, 11 are included in the list of 'Fortune 500' companies (Xi'an HTIDZ 2001). The largest of these, frequently cited by HTIDZ authorities, are Coca-Cola (a regional bottling company), Hewlett Packard, IBM, Cisco Systems and General Motors (China manufacturing facilities headquartered in Shanghai) from the U.S. Japan's companies include NEC and Mitsubishi (also in south China's Shenzhen). English companies number among them the Rolls Royce automobile company and the Swire's Group. Germany's large representatives are Siemens and Bosch. The Netherland's Philips electronics company is represented here, as well as in Shenzhen Hi-tech Park. Out of a total (mid-2002) of 3,025 enterprises in Xi'an's HTIDZ, 631 are classified as high technology companies who operate businesses or research branches on park premises. The third major State-level zone, in suburban Baoji, features companies primarily engaged in information technology, electromechanical products, composite materials from raw metals, metallic fibers, and magnetic materials. Designated high technology industry parks locate along the southern edge of the city; the industrial manufacturing zone is in the north.

Figure 10.2 Innovation Center

Following the foundation of the Xi'an High Tech Zone in March of 1991, the XIBI business incubation center was established in May 1993, 13 years after Premier Deng Xiaoping launched the opening and reform push to modernize China through boosting entrepreneurship and raising the level of its technology. National government policies to 'Develop the West' largely consist of tax reduction (in varying amounts over varying phased out time stages) and capital support from the central government. The Center occupies 475,000 sq. m., housing 248 science and technical enterprises since its inception. The range of incubating businesses covers plastics, various metal products, aviation development companies, law firms, electronics, telecommunications, several imagery companies, and several computer-related concerns as well as biomedical instrument manufacturers. Echoing the defensive military posture of its ancient fortified walls, behind which now resides a large and extensively advertised Kentucky Fried Chicken outlet owned by Taiwanese investors, the Xi'an Hi Tech Development Zone Innovation Center features trendy blue windows reminiscent of high technology buildings in the West, but shows signs of its development lag (Figure 10.2).

The Xi'an High-Tech Park now occupies 34 sq. km., with an anticipated addition of 6 sq. kilometers for an 'Industrial High Tech Area' later in 2002. In 1994 it was declared within the top four of the 53 STIP zones in China. Three years later the Xi'an HTIDZ was one of the first four zones to join the APEC industrial technology zones. Within the last four years it contributed over four per cent of the 13 per cent growth rate of Xi'an city's economy. Over 40 joint ventures also locate in the park.

University Impact

In notably great supply are Xi'an's college graduates, third highest in the country after Beijing and Shanghai. Xi'an's home province of Shaanxi (in which it is the largest metropolitan area) produces the third largest number of graduates in research and development fields, behind Beijing and Shanghai. The province boasts 2,000 science and technology research laboratories, including 50 formally recognized as being of 'advanced' standing. Of these, 55 carry the designation of National Key Laboratories at 672 research institutes. A total of 29 universities, four military universities, 66 private colleges, 126 technical schools and 15 'other' institutions of higher education enrolled 250,000 students in 2001. These produce some 26,000 annual graduates, attract 400,000 'technical labor' employees, and supply 27.5 per cent of the college-educated labor in Xi'an. Over 8,000 graduates are majors in software development alone. Xi'an high tech parks affiliated with a major university or research institute include those under development or currently operated by Xi'an Jiaotong University, Northwest Polytechnic University, Xi'an University, an Opto-Electronic Park, a New Materials Park, and a High Tech Industry Base. Labor in the park includes 54000 (half of the total) with college degrees, of which 630 hold a Ph.D. and 213 are classified as 'overseas students' (those with at least one year of higher education at a school outside China).

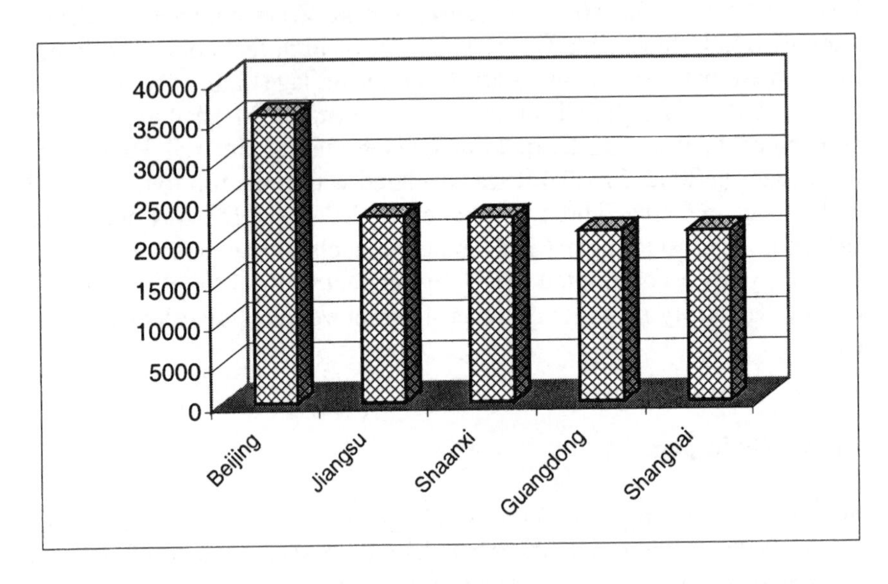

Figure 10.3 Number of scientists and engineers in major regions

XIBI Park officials point with particular pride to a Returned Overseas Chinese Students Pioneer Park. Many participants originated from local schools, pointing to the difficulty of attracting talent from other parts of China such as the more developed east coast and the pull of native roots in general. These individuals are credited with founding 140 companies since the establishment of the Returned Student incubator, a high rate for China. Most are from Shaanxi province or Xi'an itself, and typically have family members as well as roots in the region. This again resembles the profile of university graduates in the American Midwest who remain in the region to work at or start high technology related jobs rather than joining the greater migration to established regions in coastal cities (in California or Massachusetts in the U.S. example, or Shanghai, Shenzhen and Beijing in China). Some reportedly originate in Beijing or Shanghai, but come to start a business in Xi'an since it is less expensive and more supportive than the highly competitive environment in those booming, crowded cities. Most of the returned students studied in seven states in the U.S., followed by those who went to Japan, Canada, Australia, and Germany for their training. Inner China is recognized as a more challenging region for attracting both migrants and multinationals, due to its relative poverty, isolation from the outside world, and lag in development. Connections with more advanced global centers are encouraged by formal programs, however. In 1999 alone ten 'foreign specialists' offered assistance in the Center for two or three week periods. Training topics ranged from English language to 'team building' techniques (XIBI, 1999).

A variety of preferential policies exist to ease the transition of these returned students back to China, and encourage their settlement in Xi'an. If they have children, they are integrated without charge into the local school system. Entrepreneurs with a foreign passport or residence permit can start their own business with a minimum investment of US$10,000. The development zone further extends loans and equity participation to help supply needed capital. Their personal income tax is partially defrayed through various schemes, and enterprise relieved of rent payment for the first three years of their existence. Special apartment buildings are available for 'returned talents', and 'key technical and management personnel of hi technology enterprises' can receive apartments at cost. The zone will even supply a 'professional title' if it is needed but not already obtained overseas (State-Level Xi'an HTIDZ, 2000).

The new software development park displayed a typical profile for Xi'an's recently developed zones (since 2000). A large number of residences lined empty streets. Housing stock featured predominantly upscale dwellings, from colorful condominiums to freestanding 'villas'. Other amenities, presumably designed to attract high demand, mobile and high maintenance technology workers include nearby schools, parks, a decidedly upscale and expensive golf course, bonded warehouses, a broadband network for speedy and high capacity information transmission, and greenbelts that comprise four per cent of the developed land in the new section of the zone (12 sq. km.).

Five buildings house companies in the software park, arranged externally like 'fingers on a hand', considered a meaningful and auspicious arrangement. Internally, each long two-story building has an airy central hallway arcade under a glass skylight. The wide corridor includes tables and chairs along the sides, encouraging conversational flow in a relaxed setting open to anyone, a critical component of an innovation-supportive environment. Several cafeteria settings (including a fast food option as well as a sit-down choice) within each building also provide convenient places for consuming information as well as food, and building informal networks.

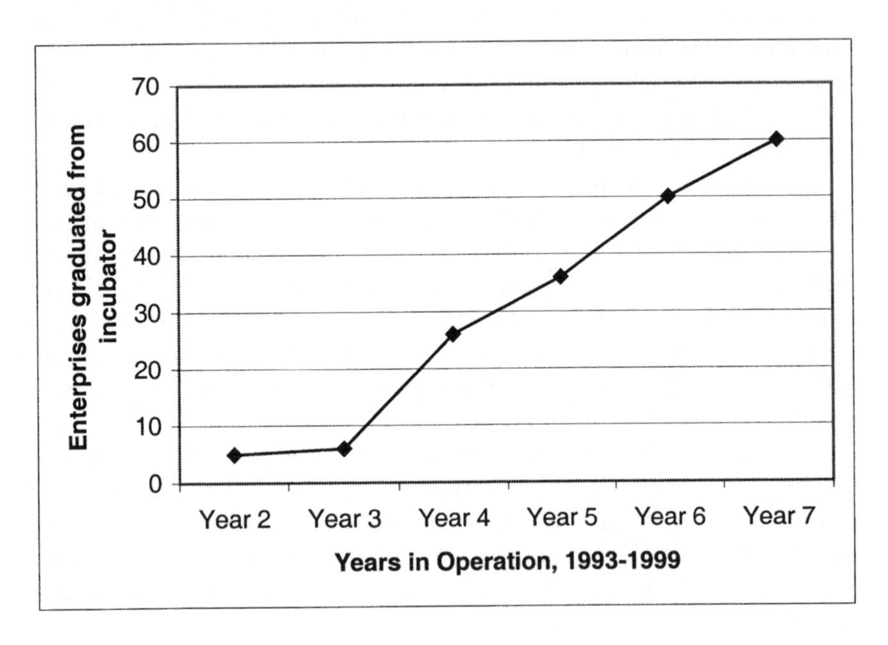

Figure 10.4 Number of enterprises graduated from Xi'an Business Incubation Center, 1994-1999

Urban Setting

Shaanxi province's principal economic draw is tourism, and Xi'an is the jumping off point for internationally well-known historical attractions noted earlier in this chapter. Within the walled old capital city, sites from dynastic China include stone steles, a time-telling bell tower (as functional now as Philadelphia's bell), and a well-established Moslem Quarter famous for its lunch foods. Modern (well-heeled) tourists can also find expensive hotels, shopping malls with imported goods, and foreign food from McDonalds to Kentucky Fried Chicken. Half-sized domestically produced taxis operate at half the start-up cost of the larger vehicles prominent in larger cities.

Outside the walled city, a vibrant modern city sprawls outward. On the southern side close to the sprouting High Tech parks, billboard signs advertise a variety of new parks as well as apartment complexes. Speculative technology parks with names like 'Blue (Silicon) Valley' and 'Chang'an High Tech Industrial Park' find support from various levels of government. The vice president of Shaanxi province was the first head of Suzhou's 'China Singapore Suzhou Industrial Park', so encourages the formation of these developments in his new posting.

Like its more established model cities, Xi'an sports a new pedestrian mall (or 'Walking Street') imitating Shanghai's Nanjing Road and Beijing's Wangfujing area. Occupied largely by shopkeepers during the workday, fountains operate down the middle at night when residents come to look at the expensive stores and restaurants. Apartment complexes recently constructed on either side supply potential patrons. A huge consumer electronics store owned by the SEG Group from Shenzhen anchors the retail strip closest to the street side, next to a tented stadium used for fashion shows and other exhibits. A new mini-downtown is also planned, complete with movie theaters and grocery stores, to cater to the southside *nouveau riche* when a critical mass is created. Jiaotong University owns and operates its own hotel, populated by business travelers as well as university-affiliated visitors.

Technology Transfer Routes

The prime functions of the university technology transfer offices include bridging university research and companies, designing projects, and acting as a technology incubator representing the university to get small and medium investors including funds to improve for global venture capital standards. They are also charged with providing protection for new intellectual property, patent, financial, and human resource requirements – a tall order under present conditions. According to interviews with several technology park and affiliated university officials, major opportunities for technology transfer include the hiring of new employees, such as students, who bring laboratory experience with them.

Companies often conduct joint research projects with each other, with universities, and teams of contracted researchers, combining fiscal capital and human talent. Many professors and students also do internships with companies, often in the form of part-time work. In other instances, companies supply laboratory equipment for conducting experiments in which they are interested. Increasingly research subjects are tailored for market needs rather than theoretical explorations. This direction follows the applied route of many universities and research centers headquartered in Xi'an for some time. Some companies bring their in-house research center. Japan's NEC software research center reportedly employs 200 people in this area, while a Singapore electronic group tests medical equipment, as a branch of a firm anchored in CSSIP from Suzhou.

Conclusion

Xi'an's setting is far more conducive to locally generated technology, and funding from other Asian interests, than to the more extensive Western involvement in Shanghai or Beijing. Profiting from encouragement by the central government, Xi'an's technology parks are designed to attract and retain locally trained talent, supplied in abundance by the numerous universities and other technology training centers, functioning as the innovation center for central China. Amenities are designed to encourage communication, including housing close to employment, recreation areas, and retail centers separate from older urban facilities. A newly engineered, modern city on a new model is developing on the outskirts of the old.

Although far from the scale of Beijing's many development districts anchored by major national universities, Xi'an presents an example of the Type III Local Innovation District on a regional scale. This makes Xi'an a more interesting study with more possibilities for duplication in other regions. The delegation from Chengdu's Sichuan University that visited Xi'an Jiaotong University two years before its STIP construction began to confer on the planning stages recognized this potential, and created their own version shortly afterward. Given the three major models illustrated in the current and preceding chapters, and China's enormous range of regions with widely varying local endowments for creating a successful STIP technopole, the final chapter discusses the possible shape of a 'Best Practice Model' with Chinese characteristics.

References

Administrative Committee of Xi'an High-Tech Industries Development Zone (July 2001), 'State-Level Xi'an High-Tech Industries Development Zone', Xi'an.

Administrative Committee of Xi'an High-Tech Industries Development Zone (2002), 'The State Level Xi'an High-Tech industry Development Zone: Investment Guide', Xi'an.

Bachman, D. (2001), 'Defense Industrialization in Guangdong', *The China Quarterly*, vol. 166, pp. 273-304.

China North West Airlines (2000).

Economist Intelligence Unit Limited (2002), *China Hand*, China.

Markusen, A. (1996), 'Sticky Places in Slippery Space: A Typology of Industrial Districts', *Economic Geography*, vol. 72, pp. 293-313.

Naughton, B. (1988), 'The Third Front: Defense Industrialization in the Chinese Interior', *The China Quarterly*, vol. 115, pp. 351-86.

National Bureau of Statistics, Ministry of Science and Technology, (ed.) (2000), *China Statistical Yearbook on Science and Technology.* Beijing: China Statistics Press.

www.chinatorch.com (2000).

www.xjtu.edu.cn/xjnet/industry (2000).

XIBI China Xi'an International Business Incubator (2000), *Xi'an Hi-Tech Industries Development Zone Business Incubation Center Annual Report 1999,* XIBI, Xi'an.

Chapter 11

A Chinese Best Practice Model

Major Issues

The central issue examined in this book focuses on the role of proximity among companies for facilitating innovation-promoting learning. Responses to the following questions posed in the first chapter will be examined in light of the empirical evidence presented, and a Best Practice model proposed:

1. Do China's science and technology industry parks (STIPs) fit previous models, or do they present a unique model with Chinese characteristics?
2. Does each STIP present its own modifications within a regional or national superstructure?
3. What is the role of proximity for promoting learning within a STIP or city cluster of high technology businesses?
4. How do interactions vary among foreign and Chinese companies, and what are the networks between foreign and domestic entities?

This chapter first re-examines relevant high technology agglomeration theories and practices based on characteristics observed in major headquarter countries. The next section summarizes empirically observed practices of STIPs and clusters in China, contrasting the models outlined in Chapter 2 with the case studies presented in Chapters 4-10. The final section proposes Best Practices possible within the Chinese setting. Given the potential for rapid change in China's political economy, and recurring fluctuations in the global business setting, this is an 'at this point in time...' analysis, with the future outcomes to be determined. Hopefully, an on-going dialogue between practitioners in the field and academic observers will yield 'lessons learned' that can lead to productive returns for all parties involved.

Western Models and Experience

Work done in North America and Western Europe on the effects of science and technology investments on economic development resulted in the following assessments by a variety of researchers looking at multiple and varied case studies throughout the most developed areas of the global economy:

1. Increased local rates of employment are associated with the presence of local excellent educational facilities, particularly in an urban location, which takes advantage of both urbanization and localization economies.
2. The economic importance of science and technology investments is attributable to the creation of skilled labor, technology diffusion, exported research, and research-intensive products (Hill and Lendel, 2002).

These two findings address important objections to science and technology investments and STIP policies in China. Some experts caution that such differential investments in specially set-aside zones increase class distinctions to favor intellectuals and business owners over agriculturalists, while further decreasing land available for critically needed food production (Wang, 2002). Unless carefully distributed throughout China, STIPs foster technopole-led growth that is inherently uneven due to the inevitable unevenness of scarce but vital resources such as investment capital, research institutions, and areas attractive to foreign companies for joint ventures. Studies conducted over a long time period in more advanced economies also indicate that, similar to the income inequality time lag shown in the Kuznets' Curve (Kuznets, 1955) where the most dangerous gap in internal development occurs just before differences level out, the future payoff in a higher standard of living for all is just a matter of time. The wide disparity in the interim can easily lead to domestic disturbances at the (albeit temporary) inequities, potentially threatening the stability of political regimes. China's response, mindful of Russia's experience, has been the proclamation of modernization with 'Chinese characteristics', meaning in this regard a willingness to make economic but not political modifications.

Numerous previous studies have shown that necessary local factors for the successful development and nurture of high technology production include:

1. An outstanding research base capable of generating innovative ideas with practical application.
2. Mediating institutions and/or individuals that can make the needed connections between locally generated ideas and businesses.
3. Adequate supportive infrastructure, from information technology to transportation.
4. Local workforce with range and depth of skills, from technical to managerial and routine support.
5. Information flows, including capital market visibility and access;
6. Locally supportive entrepreneurial culture.
7. Quality of living amenities to make location competitively attractive for individuals (Berglund and Clarke, 2000).

As predicted by market-based patterns in other countries, clusters of industry concentrations occur in China, both within and outside designated industry parks. Examples of the latter include domestic manufacture for largely domestic consumption of consumer goods such as socks in coastal Zhejiang province and cashmere in the inland city of Qinghe (Wang, 2001), and for largely export markets such as Pearl River delta's Dongguan PC industry. Examples of the former feature parks specifically configured to attract foreign investment, such as metropolitan Suzhou's cluster of Taiwanese companies (both 'dirty' traditional low wage industries and attractive high technology chip design/wafer fabricators).

Sections in each of the case study chapters detailed empirical activities in the areas of theoretical interest outlined in Chapter 2:

1. Characteristics that make a particular place attractive to a science and technology park-based economic cluster.
2. The types of firms attracted to that industrial district, and case studies of some of these firms or prominent actors.
3. How these firms in turn gave each place a particular character or niche in a domestic and/or global production chain.

4. The types of networks constructed in that place, and whether they indicate a 'learning district' or 'satellite branch' or some other mode of interaction.
5. The role of each location in integrating China into the global economy.

Chinese Realities

This section fits a summary of the Chinese STIP case studies presented in Chapters 4-10 into the overall theoretical framework outlined in Chapter 2. Given the loosely capitalist system in which businesses operate globally, what is the grounded reality 'with Chinese characteristics' in the areas examined? The schematic outlined in Figure 2.1 suggests the shape of current variations in types and functions of STIPs. Multinational companies and their Chinese subsidiaries or partners dominate the 'satellite platform'. This represents an initial, dependent stage with possibilities for being a place for labor exploitation or employment otherwise not available and replete with learning opportunities. STIPs that function as a 'hub' encompassing a variety of entities, surrounded by supplier 'spokes', comprise another model found in China. The third type of 'local innovation district' occurs in various cities as well, and holds the most promise for domestic would-be innovative ventures.

Places

Each city examined carried its own culture and locationally embedded features that inevitably affected the outcome of its STIP policies. These attributes also reflected the city's relative position to other geographic features. While general economic geography principles function similarly throughout the capitalist system, the unevenness of production factors in various national and subnational locations necessarily leads to unevenness in location choices and practices. This section first examines southern, then central and northern China locations. Each region trades on historic locational features that affect the economic outcome of parks in the various cities. The key competitive strategy for each place lies in recognizing and maximizing its place-based unique advantages.

For southern China's Shenzhen, key place characteristics are its close location to Hong Kong for financing, shipping, and intellectual connections. Other key ties are with the national language and universities of Beijing. Dongguan's inclusion in this study is less that of an STIP example (which it is not) than as an example of dynamic development at the periphery. Clearly this more western area within the Pearl River delta trades on its proximity to the vast labor pool of inner China migrants as well as its location within the Shenzhen-Guangzhou-Hong Kong axis. Its claim as a technology-affiliated area illustrates the difference between truly research-intensive and innovative high technology and the low technology, labor-intensive elements of High Tech product assembly work. Government policies for places like Dongguan that contributed to its success include the township and village enterprise structure, relaxed migration policies, and financial policies for attracting foreign investment from largely Asian sources.

Coastal central China Suzhou's attractive place characteristics trade on its attractive local natural setting within the Yangtze River delta, and - more importantly - its proximity to the bustling metropolis of Shanghai. Suzhou's small size and distance from the nearby metropolis makes it attractive for Singapore investors seeking to create a manageable showcase site, and Taiwanese companies desiring less bureaucratic oversight than in the big city. That former 'Paris of the East' is more China's New York. With a history from the 1850s to 1930s of being the most international city in China, at the mouth of the longest navigable river in the country, Shanghai is both geographically and culturally poised to be the economic powerhouse for national regeneration.

As the country's capital, north China's Beijing benefits most from the presence of the top research universities in the nation. As the most proximate site for R&D knowledge spillovers, Beijing companies pioneered the most high technology breakthroughs in the country to date. The proximity to Beijing's powerful and intrusive political organs attracts the offices of companies needing to be close to networks of power and regulations, but also discourages more innovative operations. Xi'an is also in the north, but on the eastern edge of 'inner China'. It is a regional powerhouse by virtue of its historic position as China's ancient capital and a major Third Front military-industrial outpost in the late 20th century. Consequently, Xi'an benefits like Beijing from a large number of universities specializing in applied research such as engineering. Its relative isolation means that much of its labor is locally drawn and regionally loyal. Companies are also far less overshadowed by competition from global companies than in Beijing, Shanghai, Suzhou or Shenzhen.

Firms

Firm characteristics in Chinese STIPs split more along national lines than by size. Chinese companies, which predominate in each park examined except in Suzhou, tend to be smaller and better networked than the multinationals. Potentially, they benefit more from the local incubators and university knowledge spillovers due to their inherently superior network ties with these entities. They also benefit from network ties and organizations encouraged by the government and park authorities, such as Returned Student organizations, classmate friendships, alma mater loyalties, and previous employer recommendations. Foreign firms do enter into connections with outstanding local Chinese universities, principally in order to obtain the services of potential future employees and R&D centers on university owned adjacent land. The lure remains the market potential of '...the last big market in the world – one that is roughly tripling every ten years', according to Nissan's manufacturing representative (Studwell, 2002, p.129), as well as a present source of both high and routine skill low cost labor.

Foreign firms fly the headquarter's flag but are left to navigate largely on their own in distant seas. How they cope reflects some national characteristics, including the propensity of American firms to pay taxes, employ American ethnic Chinese, and emphasize marketing and service delivery activities. German firms rely more on the exhibition halls for which they are well known in Europe, and construct in regions with large German concentrations such as Shanghai. Their predominantly small and medium size firms use these facilities for special fairs where they make supplier and other connections. They also tend to locate in more outlying areas where larger concessions can be negotiated with eager and willing small town development authorities. Taiwanese firms cluster in the two river deltas, in their own communities, negotiating the political and economic policies of the two competing regimes across the strait. Contrary to government hopes but quite understandably, foreign firms often prefer to bring their own stable of affiliated suppliers with them, firms with whom they have previously done business and developed relationships reflecting earned trust around the world. Accompanying suppliers in the wake of large relocated multinationals can number up to 50 or 60 related companies, affiliated with different operating divisions (interviews).

New Industrial Districts

The process of creating a STIP involves a transfer of land ownership or use rights. Since the State owns all agricultural land, designated district tracts need to be transferred to a development agency. A web of local permits and agreements are then negotiated, including provision of infrastructure (e.g., access to an existing expressway, or build a new one adjacent to district?). Metropolitan planning provisions, if they exist, often dissolve in the face of eagerness to obtain a development district and the bifurcation of responsibilities at the administrative level. The Department of Transportation and the Commission of Construction, for example, are responsible for building roads, though one while builds in urban areas and the other in rural areas, so room exists for duplication or competition in infrastructure provision.

Chinese STIP policy at the local level operates to create 'hub-and-spoke' districts, where the district itself operates as the large firm, seeking to attract suppliers as contingent area spokes to sustain the technopole growth localization economies for that urban area. STIPs in China are distinguished from their foreign counterparts by several prominent characteristics. The heavy involvement of the State can be seen in other developing countries, particularly those smaller than China and easier to control. The Asian 'flying geese' economic development leaders of Japan, Singapore, Taiwan, and South Korea all exhibit this trait, and thus serve as suitable examples for China's benevolent authoritarian approach. The nation-to-nation arrangement inked between Singapore and Beijing over the CSSIP in Suzhou was premised on the applicability of that model. The Suzhou park's troubles, however, came from the second major characteristic of China's development: its size and devolution of control to other players in the political-economic hierarchy.

The potential vibrancy of China's new industrial districts comes from the presence of major universities who nourish local high technology talent. The factors cited in the preceding paragraph – authoritarian State involvement – both help and harm the ability of STIPs to be learning districts for all participants. To an extent not yet reached,

Chinese authorities need to realize that their creations have reached the age of maturity, and step back so they can grow more on their own. This can occur by allowing private wealth to develop and providing incentives for its investment in companies through true venture capital consortiums and a fully functional stock market. Changes in the financial system are explored in a later section under government policy and structural recommendations.

STIP management needs to be systematized and nurtured through education of managers by best practice examples whose adoption is then encouraged throughout the country. Examples include establishing both formal and informal spaces for information exchange interactions. Another example is minimizing red tape through 'one stop' processing procedures. Creating bridges with top universities and institutions with outstanding applied research in the area benefits all participants. These untraded, non-commodified interdependencies form a vital place-based agglomeration enhancing glue, as illustrated in advanced economies that encourage knowledge spillovers. Occurrences of quasi-spin-offs in high technology industries based on R&D from affiliations with government and non-government research institution shows that this is happening on a small scale in China's most promising cities.

Science Park Networks

Interviews and surveys in the field focused on ascertaining the types of networks operating within various STIPs and their surrounding area, and the degree to which they were successful in creating a growth core within that park, city, and/or region. Two hypotheses frame this examination, based on the workings of innovative networks in the developed world. The first assumption is that information exchange is the key ingredient promoting the success of innovative enterprise. Conditions of doing business in China accentuate the importance of networked coping strategies. While transnational companies spin webs of 'relational networks' connecting their operations and far-flung places (Dicken, 2000, p. 282), their operations in any location reflect both headquarters policies and local practices to varying degrees of practical adaptation (interviews). Adaptations reflect the astuteness of the local manager, the willingness of foreign headquarters to acknowledge local realities and permit adaptations, and the challenges/opportunities presented in each place. Examples frustrating foreign investments include the lack of transparent accountability or enforcement, the quality of Chinese officials and the prevalence of bribery, turf battles between government departments, the high costs of complicated procedures and paperwork, and the challenge of obtaining local staff with foreign language skills and local connections (Nyaw, 1996; personal interviews).

The second assumption is that high technology-based businesses must be adaptable to survive in China's business environment. This research therefore looked at how, where, and what information is exchanged, both among other foreigners and with natives. While previous studies established the presence and importance of significant overseas Chinese business investments, representing approximately two-thirds of China's total foreign direct investment (Weidenbaum and Hughes, 1996; Qu and Green, 1997; Strange, 1998), the locational congruence of business and residence choices remains less examined. A key contention of this research is that while the corporation usually forms the object of examination, creation of a technologically advanced transnational community is more important to promoting international learning transfer.

Technology transfer is of two types: product and personnel. Critical technology remains largely in the home country, while production essentials may be licensed to a joint venture partner on a limited basis. Most transfer of knowledge therefore takes place by way of employee training, constituting nevertheless a not inconsiderable contribution by foreign countries to China's more technologically adept future. The difference between developed world standards of quality control, precision, processing steps, and service level is bridged by in-house training. Even for the Germans, who are known for their use of apprenticeship, training is done in China with the trip to Europe reserved for the top few employees. In more service-intensive industries such as the hotel industry, top employees spend time at a variety of overseas locales. The not uncommon practice of migrating to another company after a few years also provides an exchange of training, although those on the low end of the payroll chain seek to compensate for frequent losses by keeping training to a minimum.

University-company interactions occurred on a largely unsystematic basis. Individual examples include partnership between the combined labs of pharmaceutical giants GlaxoWellcome, SmithKline Beecham and the Shanghai Institute of Material Medica, one of China's oldest medical research institutes, in the interests of exploring combinatorial chemistry possibilities. Roche pharmaceutical recently made an arrangement with Fudan to exchange funding for a lab in return for results on a project of interest to the company in Zhangjiang.

Connections to headquarters by China-based affiliates were universally bemoaned for the perceived lack of visibility – and appreciated for the discretionary scope provided by a distant office with little interest in or understanding of the personal and business context in which their Asian affiliates operate. The difficulty communicating the differences in living and business contexts is exacerbated by the short time visiting personnel from the headquarter office spent at the foreign location, necessitating a tightly scripted and relatively smooth procession of events from a luxurious hotel base. The challenges of operating in the Chinese environment were seen as almost never officially accommodated. One example was a question about how to get reliable suppliers; where a good Chinese strategy involved enlisting competing suppliers to ensure better service at lower cost, one company representative was restricted by the company policy to relying on only a single supplier. The voluminous set of policies set by headquarters was almost universally felt best ignored, in favor of more novel approaches. A China posting usually falls to a somewhat junior or almost senior staff, anxious to rotate shortly to a more visible final position prior to retirement, thus concerned with establishing a good record for the company in a two to three year window. Shanghai functions as a way station with great risks, and opportunities, maximizing the importance of quickly establishing networks.

Shanghai's demonstrable success in attracting an increasing number of foreign companies and investments reflects both replicable structural elements and more uniquely embedded cultural elements. As predicted in previous theoretical projections of elements essential to nurturing high technology clusters, especially those heavily utilizing inputs from foreign firms, the presence of firms and residence clusters binds businesses in profit-generating proximity. Resources can be directed by government authorities to building connective infrastructure, from transportation to internet capacity, reducing bureaucratic burdens, increasing transparency and enforcement of requirements, providing housing and other amenities to create a familiar setting replicable globally: the conditions for creating sustaining networks between foreign activity nodes within the host framework.

Cultural conditions remain the elusive, ingrained, and unique element, sorting people by preference. Even in areas on similar economic footing, such as Beijing, Shanghai-Pudong, and Guangzhou-Shenzhen, distinctions remain that foreigners quickly list by identical attributes. Marketing may make a perceptual difference, as found by many Western municipalities practicing the fine art of place-product packaging, or image-making. Geographical distinctions were clearly made by all as to prevailing practices between the Beijing business environment and Shanghai. As the capital, Beijing served as the requisite location for administrative headquarters to establish a listening post for regulations and political networks. The more conservative, slower paced cultural context of the capital contributed to Shanghai's draw for the faster world of profit-loss considerations and open adjustment to foreign ways.

Creating and sustaining critical nurturing networks are a combination of design. Whether foreigners or natives occupy these networks is less important ultimately than that they are recognized as necessary, and spaces created for their support. As China solidifies its ties to the global community with WTO membership, at the same time pursuing policies in place for transferring responsibilities and jobs from foreigners to natives to soak up excess SOE labor and raise the level of local production, attention to cluster formation and network operations increases in importance.

Globalization Effects

Interview and survey responses from TNCs indicated that location choices within China for manufacturers (the great majority of companies) were constrained by two major factors: the need to be close to a joint venture partner's location (estimated to be THE deciding factor for 3/4ths of TNCs in China in the 1990s by a KPMG study) and by government directives specifying the particular location(s) within that city (Economist, 2002; personal interviews, 1999-2002). For service companies, clustering also occurred but based on another set of choices involving the location of customers, suppliers and competitors; the availability of infrastructure, labor and amenities needed; and the cost of facilities such as real estate. Chinese government policy regarding land cost acted in connection with the increased marketization of land prices to cause STIP property to skyrocket in cost. Eagerness of foreign investors to enter the China market, compounded by the cost of expatriate accommodations and benefit packages in a system that was increasingly structured to extort the most money from them as well as encourage the hiring of locals instead, led to an extremely inflated cost of doing business in the mid-1990s. Prices for land leases in STIPs in Pudong, for example, were double and triple those in major U.S. cities (Studwell, 2002).

Examples and their related best locations within China include locating in or near a major port city for companies relying on imported parts or exported product: from north to south, this means Tianjin, Shanghai, or Shenzhen/Hong Kong, for the sites studied. For those catering primarily to a domestic market, considerations can be suitable labor in a central setting on a major transportation node, which favors Shanghai but opens the possibilities for inner China and secondary cities (Chengdu, Shenyang, Qingdao) quite a bit, given reliance on a domestic sales network and national marketing. Second tier cities in peripheral regions targeted for development, such as Xi'an and Chengdu in western China, are working hard to attract desirable High Tech companies to cluster in their parks. Connections forged with local officials, provincial government, and sub-national Party powers assist in these less-favored locales that, even more so than in the larger cities, seek to build on the core competence of local firms and local specialties.

For Chinese companies in STIPs, the issue of affiliation ties (to a region or university) looms largest. The lack of suitable business training available in Chinese universities led Motorola to establish its own in-house facility, dubbed 'Motorola University'. Business practices taught range from management skills and accounting practices to encouraging employees to speak their own mind and contribute ideas for improving business practices and products (interview, June 2002). Responding to China's desire for more training of Chinese employees, General Motors established a GM Technology Institute at Qinghua University (Beijing) and Jiaotong University (Shanghai), China's premier engineering colleges (Studwell, 2002).

Suitable and scarce labor is a major concern for high technology companies, which favors the 'bright lights' cities of Beijing and Shanghai. Shenzhen is relatively more of a draw for Chinese firms, and much less so for foreign companies. The configuration of services offered at various STIPs can make a difference, but many infrastructure considerations are becoming standard, such as high speed telecommunications connections, a range of good housing from worker affordable to manager amenities, recreation and education outlets, public transportation connections, management assistance with tax and other bureaucracy-navigating issues. Areas outside nationally designated zones and major cities attract businesses looking for cost-cutting locations close enough to urban areas, such as Dongguan, Suzhou and Taicang (outside Shanghai).

STIPs located on the outskirts of their cities (with room to grow at reasonable rates) in proximity to major universities such as in Beijing, Shanghai and Xi'an, are the most promising for domestic high technology companies and foreign ventures alike. Taking advantage of the knowledge spillover proximity effect nurturing High Tech businesses, China's Ministry of Education and Ministry of Science and Technology aim to create 100 college High Tech parks by the year 2010 (www.china.com.cn, 2002). More truly low-tech enterprises, relying principally on low cost labor and the round-trip 'exportation' of products to appear compliant with Chinese requirements, cluster in other STIPs that provide tax havens and a relatively prestigious address. Conditions in other designated special areas such as ETDZs, Free Trade Zones, and those established by non-national government entities, organizations, or private sources were not examined in this research since they are not areas of significant high technology or foreign-native corporate location. The 'big fish, small pond' effect can be significant for individual companies, however, in negotiating concessionary terms with local property providers and authorities eager to establish their reputation with a big catch.

China's system of STIPs and development zones is evolving unevenly, partially by design to favor certain regions - first Beijing and the Pearl River delta, then metropolitan Shanghai and most recently western inner China, in a revolving balancing attempt. China's acceptance into the World Trade Organization counts as yet another step in its imbrications into the global financial system. Full membership, with the attendant risks and rewards, necessitates following generally accepted policies and practices, in which China lags in several important respects. The following concluding section outlines areas where China could and should bring its system into closer compatibility with the best practices of top players in the field in which it seeks to play, finding prosperity by becoming a more adept neo-Marshallian node nurturing foreign and domestic technology clusters in a recognizable landscape.

Challenges

Problems impeding innovation commodification in China include:

- Obtaining venture capital in a risk-averse, conservative society that also restricts foreign financial access.
- A thin knowledge base for applied research, that needs to master the old foundation in order for new products to mature, rather than rushing to market insufficiently tested and grounded products.
- Lack of knowledge in Chinese firms of how to efficiently and at a sufficiently high quality level manufacture and market products. Firms have little business experience and few good teachers.
- Need for affordable space close to R&D institutions, enabling easy access to professors, laboratories, and other institutional support. In-house corporate research is seldom done due both to initial expense and the high risk of leaks.

STIPs are therefore needed to pool thin resources, provide visibility for new efforts to target assistance to foreign and domestic firms. Clusters naturally occur where needed attributes are located. The question is, how to provide identifiable attributes? A major geographic observation lies in the unevenness of places in their natural and human endowments. Some systemic provisions can even the field within that framework, providing national advantages within a global system. Places can then market their unique endowments for their appropriate niche. Policies for China to accentuate its strengths and meet its challenges are outlined in the final section.

Best Practice Possibilities: Policy Initiatives and Market Institutions

Regulatory Framework and Transparency

Promulgating more regulations to *protect intellectual property* and enforcing them through a well-informed and sufficiently large judicial system is a key recommendation of many foreign businesses. China will not experience the quality and quantity of technology transfer that it desires domestically without great improvement in the current situation. This is a particular problem for the pharmaceutical industry, which China claims to target for attraction given its large aging population. Increasingly strident promotion of native remedies can be seen as an admission that people prefer to balance traditional remedies with the more expensive and reliable foreign formulations. Chinese pharmaceutical companies reputedly face lighter quality enforcement provisions, and benefit from a rampant lack of intellectual property protection in this field. Stricter enforcement of quality standards would benefit Chinese companies who legitimately engage in biotechnology research and joint development efforts with companies in more developed countries, but whose products are not valued by the purchasing public with the same confidence as those of Western counterparts.

Fiscal System

Opening of capital markets to permit loans by local subsidiaries of foreign banks would greatly increase the amount of capital available to businesses in China. At present, only financial institutions with China headquarters in the Liujiazui district of Pudong are permitted to begin to engage in limited transactions of this sort. The state of China's banks are widely seen as a major weakness of the present business structure, threatening to undermine the entire economy with their 'triangular debt' system of unpaid loans transmitted to SOEs and weak banks, backed by an undermined national bank. Permitting the development of a credible, well-managed stock market would offer an alternative to these loans as a source of capital, freeing uninvested savings and permitting the market to weed out or reward companies by their performance.

Permitting the *accumulation of private wealth and incentivizing its investment via venture capital* in domestic companies would *encourage domestic entrepreneurs* and commodifiable innovations. Successful business practices should be taught more aggressively, with a more hands-on approach to nurturing incubating and fledgling ventures. Fiscal and tax incentives for foreign firms should be extended, and enforced broadly and consistently in order to be meaningful. This is not currently the case; the ensuing leakage of revenue feeds systemic corruption and impoverishes the nation needlessly.

Several steps could be taken to encourage foreign companies *to invest in the interior provinces*. Inducements presently advocated by the central government are largely limited to the standard tax deferment that, in light of the purported infrequency of payment, adds little additional attraction to companies. The Organization of Economic Cooperation and Development (OECD 2002) suggests that more transparent regulations would be particularly appealing in the inland areas, where distance from the central authorities encourages local officials to be more 'innovative' in their regulations and uneven enforcement.

Alleviating fiscal burdens on local governments, which are particularly burdensome in the more rural and smaller urban areas, would also free capital for business-friendly infrastructure improvements. As the larger cities increasingly grow by absorbing small towns in the surrounding peri-urban area, they also integrate new enterprises that have been growing in these border regions. Some foreign companies would miss the autonomy and creative dependency of these smaller areas, however, that are now so eager to attract businesses (especially the relatively capital rich foreign companies) to their STIPs that they will oblige these companies with services and infrastructure to a degree that would be more difficult to negotiate with their bigger neighboring city. Best Practice would regularize these conditions, however.

Labor Policy

Accelerated *removal of impediments to labor mobility*, such as further lifting of the *hukou* requirements, would be a business-friendly move for manufacturers. It also might cause an exodus to the cities that China has so far controlled much better than other developing countries, choosing to take a middle path by largely tolerating millions of 'floating population' undocumented workers who supply construction labor for rapidly growing urban areas. Regularizing this movement would be better for workers since it would extend services now denied (health care, education for their offspring) and provide a better work force for employers as well.

Infrastructure Connections

Proceeding with plans to provide *more deep sea ports* for accommodating modern vessels is part of the trade promotion program and would even the advantages and choices between north, central, and south coastal China. *Linking all major productive regions of the country with shipping facilities* such as rail and truck transport is also crucial for diversifying business location choices and spreading employment opportunities. Moves in this direction in second tier cities presently hold exciting possibilities for future accelerated development in more diverse regions.

Technology Transfer

China has made enormous changes in its educational system, responding to a variety of pressures. More needs to be proactively planned so that both higher education and its economic applications can 'walk on two feet' without neglecting the missions of instruction, basic research, and applied transfer of research to the marketplace to enrich both originator and the general public. In the short term, restructuring of China's research bodies and reward structure in an effort to emphasize applied research allowed older inventions (and inventors) to come to the market, but downplayed the importance of both teaching and theoretical research. The tilt in a top-down, resource limited system produced 'big heads, weak legs' for innovations to stand on as a result. The lure of multinational companies offering large salaries and benefit packages to advanced students and recent graduates decreases China's ability to produce its own intellectual property; the lack of risk capital for start-up companies hampers these young employees chances of starting their own enterprises as well.

Some best practice procedures adopted in more advanced countries include encouraging student and professorial *internships in companies, and the use of university facilities by companies*, to create greater partnership and understanding between these two cultures. *University technology parks* promote the initial overhead-free exploration of entrepreneurial possibilities. STIPs that are removed from the campus but in urban proximity promote the next stages through subsidizes incubators and affordable office spaces. Services go beyond hardware (building space, computers, office furnishings) to critical training through providing advice and opportunities for trust building in informal settings and information exchange in all settings.

Countrywide reforms must *focus on the total system and the interaction of all relevant parts*, as outlined above. The health of the whole grows only as each segment strengthens. While some support is necessary in a transitioning political economy with greater needs than resources at present, each segment should be assisted but allowed to learn through making mistakes and improving by coming up with new remedies suitable for its particular circumstances. Institutional rigidities should continue to yield to pragmatic adjustments permitting openness and learning within China's evolving political economic system.

STIP managers frequently commented on the comparative freedom to innovative in cities with the greater distance from administrative centers and the correspondingly least number of competing ranked officials whose favor and cooperation must be sought. Beijing, the national capital, is pitied as the most rigid and burdened with a plethora of high-ranking bureaucrats. Pudong STIP officials at least have the Huangpu River between them and Puxi's metropolitan mandarins. Shenzhen is at a blessedly greater distance from Guangzhou, the region's urban giant, while Xi'an can enjoy the fact that the 'emperor is far away' from the western provinces. One of the hardest but most necessary steps in the transition to a freer, more innovative and market-responsive economy remains the need for central authorities to permit flexibility within the same regulations, while promoting transparency and equal enforcement.

China's challenges are greater than publicly acknowledged in the picture presented by available statistics – but the possibilities for the future remain bright. Science and technology industrial parks constitute critical space for enhancing economically related information flows among Chinese and foreign headquartered companies, and (less seldom) between the two. China correctly proceeds down two tracks as it utilizes these privileged spaces to both learn from more advanced country companies and to advance its own applied research commodification in the (so far successful) quest to become a global competitor. The process well merits ongoing observation, as the outcome is critically important for the long-term stability of the global economy.

References

Berglund, T. and Clarke, M. (2000), *Using Research and Development to Grow State Economies'*, National Governors' Association, Washington, D.C.

Dicken, P. (2000), 'Places and Flows: Situating International Investment', in Clark, G. Gertler, M. and Feldman, F. (eds.), *Handbook of Economic Geography*, pp. 275-91, Oxford University Press, Oxford, UK.

Economist Intelligence Unit Limited (2002), *China Hand*. The Economist.

Hill, E. and Lendel, I. (2002), 'The Impact of the Reputation of Bio-life Science and Engineering Doctoral Programs on Regional Economic Development', Paper presented at Georgia State University.

Kuznets, S. (1955), 'Economic growth and income inequality', *American Economic Review*, vol. 45, pp. 1-28.

Nyaw, M. (1996), 'Investment environment perceptions of overseas investors of foreign-funded industrial firms', in Yung, Y. and Sung, Y., *Shanghai: Transformation and Modernization Under China's Open Policy*, pp. 250-72, Chinese University Press, Hong Kong.

Organization for Economic Co-operation and Development (2002), 'Synthesis Report', *China in the World Economy: The Domestic Policy Challenges*, OECD, Paris.

Qu, T. and Green, M. (1997), *Chinese Foreign Direct Investment: A Subnational Perspective on Location,* Ashgate, Aldershot, UK.

Strange, R., (ed.) (1998), *Management in China: The Experience of Foreign Businesses*, Frank Cass, London.

Studwell, J. (2002), *The China Dream: The Quest for the Last Great Untapped Market on Earth*, Atlantic Monthly Press, New York.

Wang, J. (2002), 'High and New Technology Industrial Development Zones', in Webber, C.M., Wang, M. and Zhu, Y. (eds.) *China's Transition to a Global Economy,* Palgrave Macmillan Global Academic Publishing, London.

Weidenbaum, M. and Hughes, S. (1996), *The Bamboo Network*, The Free Press, New York.

www.china.com.cn/english/873.htm (10/29/2002), 'China to set up 100 college High Tech parks', from *Huasheng Monthly.*

Index